THE **MARINE BIOLOGY**
COLORING BOOK

The
MARINE
BIOLOGY
COLORING
BOOK

by Thomas M. Niesen

Illustrations by
Wynn Kapit and
Lauren Hanson

HarperPerennial
A Division of HarperCollins*Publishers*

Thomas Niesen is an Associate
Professor of Marine Biology at San
Francisco State University. He teaches
courses and conducts research in
marine biology and ecology. His hobbies
are diving, photography and gardening.

This book was produced by Coloring
  Concepts Inc.
P.O. Box 324, Oakville, Ca. 94562

The book editors were Joan Elson, Carol
  Denison and Larry Elson
Consulting editor was Dr. Gary Brusca
The copy editor was David Cross
Type was set by Ampersand Design,
  San Francisco
Page makeup and production coordination
  was by Donna Davis
The proofreader was Sue Gamlen

ISBN 0-06-460303-2

97 98 99   23

DEDICATION

To my friends Jack, Carolyn, Marietta, Ken and Maryrose, and Matt and Anne for their encouragement to write this book.

To my friend and collaborator Wynn Kapit for making this effort such an enriching experience.

To my wife, Anne, and daughter, Amy, for their constant love and understanding which made this book a reality.

PRODUCER'S ACKNOWLEDGMENT

Coloring Concepts had tremendous support during the development of this and the other three books of this initial series (Botany, Human Evolution, Marine Biology, and Zoology). The people at Harper and Row, especially Irv Levey and Tom Dorsaneo, have encouraged us, responded to our needs, and demonstrated patience and understanding in our struggle to make a deadline. The following friends have directly aided our endeavor: Howard Nemerovski, Tom Larsen, Stu Boynton, Kathy Dahl, Bill and Danielle Brown, Don Jones, Terry Anderlini, John Moran, and Dr. Jack Lange. Their support and confidence in us is greatly appreciated. Dr. Lange (Lange Medical Publications) has given us advice and direction as we develop as publishers. His philosophy of publishing has been a source of inspiration to us. We are much indebted to Donna Davis, who coordinated our book productions, and the fine people who worked with her: Libbie Schock and Nancy Steele at Ampersand Design, Sue Gamlen, Bob Nemerovski, and our copy editor, David Cross. The concern and reassurance of our friends Gene Mattingly, Regan Anderlini, Ken and Ute Christensen, Ellis and Merilyn Bowman, Maurice and Ann MacColl, Bob and Nancy Teasdale, Sharon Boldt, and Julie and Nancy at Schoonmaker provided the background so necessary for us to achieve our goal. We are very grateful.

July 1981

Joan Warrington Elson
Lawrence M. Elson

TABLE OF
CONTENTS

PREFACE

The sea has always been a place of mystery and wonder. Stretching beyond the horizon and teeming with a bewildering variety of creatures, the sea has appeared incomprehensible to us. The pioneering explorations of the oceans continue to be portrayed as attempts to probe the ocean's "mysteries" and unlock its "secrets." This portrayal certainly captures our attention but at the same time we are made to feel incapable of ever understanding life in the sea.

Marine biology, the study of the sea's living organisms, is accessible and enjoyable. In this book I introduce the diversity of marine plants and animals to demystify marine biology and yet retain the feeling of wonder and respect for marine life. This book utilizes a unique educational tool that allows the reader to interact with the written text by coloring. By following the straightforward coloring directions, it is possible to very pleasantly create your own visual interpretations of marine life. Using the same color for a structure common to several organisms allows you to understand both the similarity and diversity of form and function in marine organisms.

The text is organized into several segments, each comprising a series of self-contained subjects with corresponding plates to be colored. Each subject and plate can be read and colored independently, although completing an entire segment in sequence would probably be more meaningful. Appropriate cross references (by plate number) are made throughout the book. Segments of the book introduce the major marine habitats, present the diversity of marine plants and animals, trace important life cycles, and explore the complex interrelationships among marine organisms. To emphasize the universality of marine habitats in all the oceans, no single geographic locality is represented exclusively. Rather, organisms from throughout the world are discussed and illustrated.

The author is indebted to a number of people who aided in the creation of this book. Their assistance was of inestimable value and is deeply appreciated. The energy and vision of Larry Elson and the encouragement and thoughtfulness of Joan Elson are gratefully acknowledged. The insight, creativity, and consummate artistic skills of Wynn Kapit are evident throughout this book, and I thank him for his guidance and kindness. The beautiful art work of Laurie Hanson is sincerely appreciated. The reviews and criticisms of many colleagues are thankfully acknowledged, especially Drs. Margaret Bradbury, Michael Josselyn, William Neff, Gary Brusca, and the staff of the California Academy of Sciences including Dalene Reid, Dr. Daphne Dunn, Dr. Welton Lee, Barbara Weitbrecht, Vincent Lee, Dustin Chivers, Jacqueline Schonewald, and Tomio Iwamoto. I thank Joan and Larry Elson, Carol Denison, Dr. Gary Brusca, and David Cross for their tireless editorial contributions. And, finally, I thank my wife Anne for shouldering the responsibility of keeping body and soul together while I wrote this book.

Thomas M. Niesen

July 1981
Half Moon Bay, Ca.

HOW TO USE THIS BOOK
COLORING INSTRUCTIONS

1. This is a book of illustrations (plates) and related text pages in which you (the colorer) color each structure indicated the same color as its name (title), both of which are linked by identical numbers (subscripts). In the doing of this, you will be able to relate identically colored name and structure at a glance. Structural relationships become apparent as visual orientation is developed. These insights, plus the opportunity to display a number of colors in a visually pleasing pattern, provide a rewarding learning experience.

2. You will need coloring instruments. Colored pencils or colored felt-tip pens are recommended. Laundry markers (with waterproof colors) and crayons are not recommended: the former because they stain through the paper, and the latter because they are coarse, messy and produce unnatural colors.

3. The organization of illustrations and text is based on the author's overall perspective of the subject and may follow, in some instances, the order of presentation of a formal course of instruction on the subject. To achieve maximum benefit of instruction, you should color the plates in the order presented, at least within each group or section. Some plates may seem intimidating at first glance, even after reviewing the coloring notes and instructions. However, once you begin coloring the plate in order of presentation of titles and reading the text, the illustrations will begin to have meaning and relationships of different parts will become clear.

4. As you come to each plate, look over the entire illustration(s) and note the arrangement and order of titles. Count the number of subscripts to find the number of colors you will need. Then scan the coloring instructions (printed in boldface type) for further guidance. Be sure to color in the order given by the instructions. Most of the time this means starting at the top of the plate with A and coloring

in alphabetical order. Contemplate a number of color arrangements before starting. In some cases, you may want to color related forms with different shades of the same color; in other cases, contrast is desirable. In cases where a natural appearance is desirable, the coloring instructions may guide you or you may choose colors based on you own knowledge and observations. One of the most important considerations is to link the structure and its title (printed in large outline or blank letters) with the same color. If the structure to be colored has parts taking several colors, you might color its title as a mosaic of the same colors. It is recommended that you color the title first and then its related structure. If the identifying subscript lies within the structure to be colored and is obscured by the color used, you may have trouble finding its related title unless you colored it first.

5. In some cases, a plate of illustrations will require more colors than you have in your possession. Forced to use a color twice or thrice on the same plate, you must take care to prevent confusion in identification and review by employing them on separate areas well away from one another. On occasion, you may be asked to use colors on a plate that were used for the same structure on a previous related plate. In this case, color their titles first regardless of where they appear on the plate. Then go back to the top of the title list and begin coloring in the usual sequence. In this way, you will be prevented from using a color already specified for another structure.

6. Symbols used throughout the book are explained below. Once you understand and master the mechanics, you will find room for considerable creativity in coloring each plate. Now turn to any plate and note:

a. Areas to be colored are separated from adjacent areas by heavy outlines. Lighter lines represent background, suggest texture, or define form and (in the absence of "don't color" symbols) should be colored over. If the colors you used are light enough, these texture lines may show through, in which case you may wish to draw darker or heavier over these lines to add a three-dimensional effect. Some boundaries between coloring zones may be represented by a dot or two or dotted lines. These represent a division of names or titles and indicate that an actual structural boundary may not exist or, at best, is not clearly visible.

b. As a general rule, large areas should be colored with light colors and dark colors should be used for small areas. Take care with very dark colors: they obscure detail, identifying subscripts, and texture lines or stippling. In some cases, a structure will be identified by two subscripts (e.g., A + D). This indicates you are looking at one structure overlying another. In this case, two light colors are recommended for coloring the two overlapping structures.

c. Any outline-lettered word followed by a small capitalized letter (subscript) should be colored. In most cases, there will be a related structure or area to color. If not, the symbol N.S. (not shown) will follow the word; or, the word functions as a heading or subheading and is colored black (●) or gray (★). Outline titles (headings) with no subscript following are to be left uncolored.

d. In the event structures are duplicated on a plate, as in left and right parts, branches, or serial (segmented) parts, only one may be labeled with a subscript. Without boundary restrictions or instructions to the contrary, these like structures should all be given the same color.

e. In looking over a number of plates, you will see some of the following symbols:

●	= color black; generally reserved for headings/subheadings
★	= color gray; generally reserved for headings/subheadings
-¦-	= do not color
A( )	= set next to titles subscript; signals this structure composed of parts listed below with same letter but different exponents; receives same color; only its parts are labeled in illustration
A^1, A^2, etc.	= identical letter with different exponents implies parts so labeled are sufficiently related to receive same color
N.S.	= not shown

7. In the text, certain words are set in *italics*. According to convention, the generic name and species of an animal or plant are set this way (e.g. *Homo sapiens*). In addition, the title of any structure to be colored on the related (facing) plate is set in italics (except for headings and subheadings). This is to enable you to quickly spot in the text the title of a structure to be colored.

THE **MARINE BIOLOGY**
COLORING BOOK

1
TIDES

The ocean is a huge body of salt water that covers nearly three-quarters of the earth. Its surface is moved by wind and wave, and it is in constant motion, flowing in slow, oceanwide currents.

Tidal variations are caused by the interplay of the *earth, sun,* and *moon.* As the *moon* travels around the *earth,* and as they both, together, complete a larger path around the *sun,* it is their combined gravitational attraction together with other factors that causes the alternate rising and falling of the surface of the ocean, called tides.

Begin coloring on the diagram at the upper left, which depicts the earth, its orbit around the sun, and its rotation on its axis; and the moon and its orbit.

Gravitation (gravity) is an attraction between two masses; it is proportional to the product of their masses and inversely proportional to the square of the distance between them. Thus, although the *moon* is much smaller than the *sun,* it is so much closer to the *earth* that it plays the major role in causing tidal variations.

Visualize an *earth* completely covered with a layer of water, and spinning on its axis. As the *moon orbits* around the *earth,* that part of *earth* directly opposite the *moon* will be pulled by the *moon's gravity,* and the water will *bulge* away from the *earth* toward the *moon.*

The spinning motion of the earth–moon pair creates what is known as a *centrifugal force.* This force causes the water on the side of the earth farthest away from the *moon* to also *bulge* out, away from the earth's surface. The *gravitational pull* of the *moon* and the *centrifugal force* modify one another, and, at any given time, a *bulge* of water will occur on the side of the *earth* facing the *moon* and on the side farthest away from the *moon.*

Color the three drawings of the earth on the right. The first (top) drawing illustrates the moon's gravitational pull on the earth's surface water. The second drawing illustrates the tide caused by the centrifugal force, and the third indicates the combined effects of the two forces. Color the word tidal a B color and the word bulge a D color.

Because the *earth* revolves completely around on its axis approximately (but not precisely) every 24 hours, every spot on the *earth* will experience both types of *bulges* within that 24-hour period. The *bulges* are called the lunar tides: the peaks are the high tides, and the troughs (between the bulges) are the low tides.

In an isolated earth–moon system, each spot on the *earth* would experience four tides each lunar day (24 hours 50 minutes): two high tides and two low tides of equal magnitude. But the *sun* has an important modifying effect.

Color the diagram in the center of the page that demonstrates the position of the earth, sun, and moon during the moon's phases, and illustrates the tidal bulges. Then color the tidal range, as indicated on the drawing in the lower left corner, as well as the map that shows tidal variations.

The *moon orbits* the *earth* once every 27.5 days (the lunar month), and in the course of its *orbit* the *moon* is in a different position relative to the *sun* every day. At the time of new moon and full moon, the *sun, earth,* and *moon* are in a direct line with one another, and the combined *gravitational pull* causes extra high and extra low tides. These tides are called *spring tides* and are depicted in the *tidal range* illustration to the left of the piling.

During the first-quarter and third-quarter phases of the *moon,* the three masses of *earth, sun,* and *moon* are not in direct line with one another. The *sun's gravitational pull* works to minimize the effects of the *moon's gravitational pull,* and the tidal variations are of much smaller magnitude. These tides are called *neap tides* and are depicted in the tidal range illustration to the right of the piling.

There is considerable tidal variation on earth. The geographical position, shape of the ocean basin, and a host of other local, global, and planetary factors act to modify the tides. The tides at the end of a long inlet will be considerably amplified due to the large volume of water pressed into a very narrow space, whereas the tides on a mid-ocean island will hardly be noticed.

The frequency of the tides also varies from place to place. The coastline of northern Europe experiences two equal high and low tides every day. The coast of California also experiences two high and two low tides daily, but they are not equal. The Gulf of Mexico has only one high and one low tide each day.

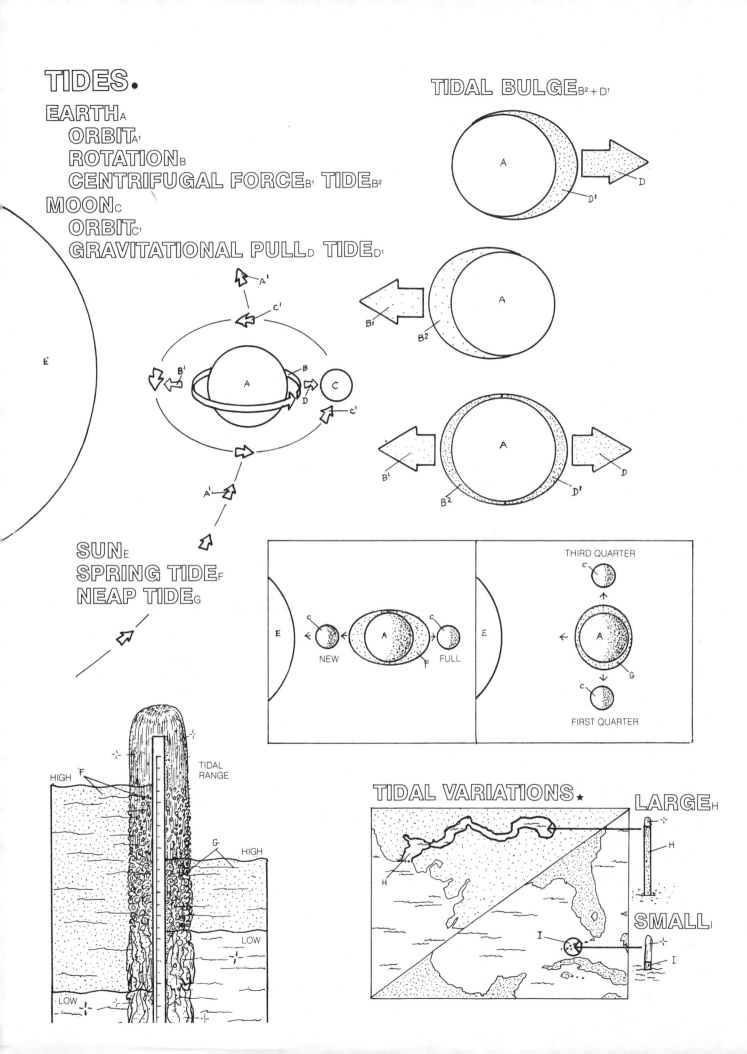

TIDES.

EARTH$_A$
   ORBIT$_{A^1}$
    ROTATION$_B$
     CENTRIFUGAL FORCE$_{B^1}$ TIDE$_{B^2}$
MOON$_C$
   ORBIT$_{C^1}$
    GRAVITATIONAL PULL$_D$ TIDE$_{D^1}$

TIDAL BULGE$_{B^2 + D^1}$

SUN$_E$
SPRING TIDE$_F$
NEAP TIDE$_G$

TIDAL RANGE

HIGH

HIGH

LOW

LOW

THIRD QUARTER

FIRST QUARTER

NEW

FULL

TIDAL VARIATIONS ★

LARGE$_H$

SMALL$_I$

2
TIDAL ZONATION PATTERNS

The intertidal zone, often called the littoral zone, is the most readily accessible of marine habitats. The intertidal zone, as the name implies, is that portion of the ocean shore that is periodically covered by the highest spring tides and exposed by the lowest spring tides. It is a very complex habitat that supports a great diversity of animals and plants. It is the meeting place of the marine and terrestrial (land) environments and is exposed—at least a portion of it—to both air and water elements during every tidal cycle. During high tide, the water temperature is relatively even, though the intertidal inhabitants may be pounded by waves and wave-borne debris. When the tide is out (low tide), the intertidal zone is exposed to extremes in temperature, variations in light, and to fresh water; snow and ice may occur in the intertidal zone in some areas.

The living conditions for resident organisms vary within the intertidal zone. Conditions are more terrestrial in the upper intertidal, and correspondingly more marine in the lower intertidal area. The degree to which a given area is terrestrial or marine is governed by the amount of time each area is exposed to the air and the frequency with which such exposures occur. This wide range of environmental conditions influences the kinds of organisms that can live at a particular intertidal level. The zonation patterns that correspond to particular tidal levels are created, in part, by this range of physical conditions. Other factors, including the presence of certain organisms, also influence zonation, and these will be explored later.

The intertidal zonation phenomenon is most visible on protected shores, where, at low tide, distinct horizontal stripes of color and/or texture are apparent. These stripes or zones take the color of the most common and/or most visible organisms present. There is a remarkable similarity in rocky intertidal zonation patterns around the world. After studying world intertidal zonation for thirty years, Ann and T.A. Stephenson devised a general scheme by which to describe this zonation. Using their scheme, we will look at the zonation pattern found at one rocky intertidal area (in Coos Bay, Oregon), with the understanding that this scheme could be applied almost anywhere in the world, although it would include different organisms.

A representative dominant resident of each tidal zone is pictured on the right side of the plate. As each zone is introduced in the text, color the resident and its corresponding zone the same color. If you wish to use realistic colors, use a light green for A; B, gray; C, light gray; D, light brown; and E, brown.

The *supralittoral zone* is the area above the high tide mark that receives both wave splash and sea-water mist. Here live terrestrial organisms, such as lichens, that can tolerate some sea water, and marine animals that are becoming less dependent on, or less tolerant of, the ocean than those living lower in the intertidal. (An example is the large isopod in Plate 28.) The small *green alga* (seaweed), is found in the *supralittoral zone:* fresh water seeping down the cliff face and sea water splashing upwards provide conditions uniquely suited to this plant.

Below the *supralittoral zone* is the *supralittoral fringe* or "splash zone." This is the upper level of the high tide zone and receives a regular splash of waves when the tide is in. Here, the marine *periwinkle* snail is found. It can tolerate long periods of exposure to air and needs only an occasional wetting.

The lower limit of the *supralittoral fringe* is marked by the beginning of a *barnacle* zone. This area is called the *midlittoral zone* and encompasses the majority of the intertidal area. It extends down to the upper limits of the habitat of the large *brown algae.* The *midlittoral zone* supports a great variety of marine animals, including mussels (not shown).

The brown alga, *Alaria,* marks the *infralittoral fringe,* which includes the lowest level exposed by extreme spring tides. This area is often occupied by the brown alga, *Laminaria,* which extends into the *infralittoral fringe* and *infralittoral zone,* or subtidal area, marking the beginning of the marine environment that is below the tides. Sponges, sea urchins, and abalone are also found in the *infralittoral zone* and *fringe.*

Low spring tides reveal most of the shore and are the best times to venture into the intertidal zone.

TIDAL ZONATION PATTERNS.

SUPRALITTORAL ZONE_A
  GREEN ALGA_{A¹}
SUPRALITTORAL FRINGE_B
  PERIWINKLE_{B¹}
MIDLITTORAL ZONE_C
  BARNACLES_{C¹}
INFRALITTORAL FRINGE_D
  BROWN ALGA (ALARIA)_{D¹}
INFRALITTORAL ZONE_E
  BROWN ALGA (LAMINARIA)_{E¹}

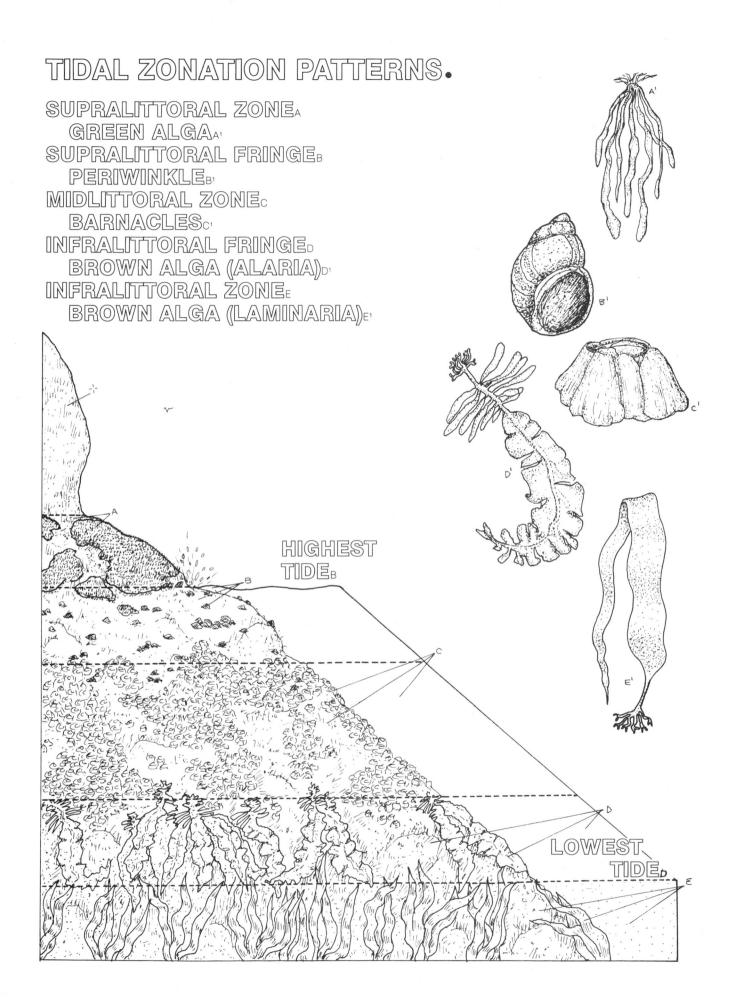

HIGHEST
TIDE_B

LOWEST
TIDE_D

3
CHARACTERIZATION: ROCKY SHORES

The shoreline where the land meets the ocean is a product of the earth's geological history, whose diverse twists and turns are often visible in the scoured rocks of the intertidal. Some rocky intertidal zones are relatively flat with little surface relief, while others are steep or irregular with boulders, ledges, overhangs, and tidal (surge) channels. Irregular and varied surfaces offer a great deal of potential living space for organisms.

All open-coast rocky intertidal areas are subjected to the battering of waves as well as being periodically exposed to the air. Using this plate we will investigate how some of the more common rocky intertidal animals are adapted for survival in this rigorous habitat.

Begin your coloring with the limpets (A) near the top of the rocks, and color each animal as it is mentioned in the text. The rocky substratum and the unidentified plants should not be colored. Note that the tide is low in this illustration and that the approximate levels of high and low tide are indicated. The bottom third of the page, from the water line on down, may be colored over with a light blue or light green.

Limpets are commonly found in the high intertidal zone. These small gastropods are able to hold onto the substratum with a large muscular foot. They seek out depressions and ledges that shield them from direct sunlight; this helps them to avoid being dried out during low tide. *Limpets* also secrete a sticky mucus that seals their shells to the rocky surface and prevents the evaporation of water from their soft bodies. Some species use their radulas (a file-like structure; see Plate 89) to excavate a depression in the rock. These *limpets* fit into the hole perfectly during low tide. When high tide returns, they move about on the rocks, grazing on algae, and then return "home" to their self-made, shallow "fox hole."

The purple *shore crab* has a flattened body that allows it to slide under and between rocks to avoid exposure. This nimble scavenger can live high in the intertidal zone. It is most active at night and at high tide.

Mussels are commonly found in the middle intertidal zone, often in large numbers. This bivalve mollusc anchors itself to the rocks or to other *mussels*. By keeping its shell tightly closed, the *mussel* is able to resist desiccation (dehydration) during low tide. A large grouping of *mussels,* called a *mussel clump,* provides living space for many other small marine creatures (not shown).

Another organism found abundantly in the high and middle intertidal region is the *aggregating anemone.* This small *anemone* (2.5 cm across) forms masses comprising hundreds of individuals. The crowding reduces water loss caused by exposure. The bodies of these *anemones* are often covered with small bits of shell and rock, which further protects them from direct exposure.

An algae-eater of the middle intertidal zone is the *mossy chiton* (ky-ton). This mollusc's shell is divided into eight separate plates, or valves, that move against one another (articulate) allowing the *chiton* to conform to the irregular rocky surface. It uses its foot to maintain a strong grip against the wash of the waves. When the environment is calm, the *mossy chiton* forages on small algae, and remains on the rock's surface during both high and low tide.

In the lower intertidal zone, there are often sharp ledges and overhangs where light cannot penetrate and the growth of algae is prevented. *Sponges* and other delicate animals such as the ostrich-plume *hydroid* attach themselves to these protected underledges.

The purple *sea urchin* lives in both the lower intertidal zone and below, in the subtidal zone. It survives the pounding of waves by excavating a pit-like depression in the rock with its spines. The pits remain moist at low tide and protect the *urchins* from the direct crashing of waves at high tide.

ROCKY SHORES.

HIGH TIDE ★ →

LIMPET_A

SHORE CRAB_B

MUSSEL CLUMP_C

AGGREGATING ANEMONE_D

MOSSY CHITON_E

LOW TIDE ★

SPONGE_F

HYDROID_G

SEA URCHIN_H

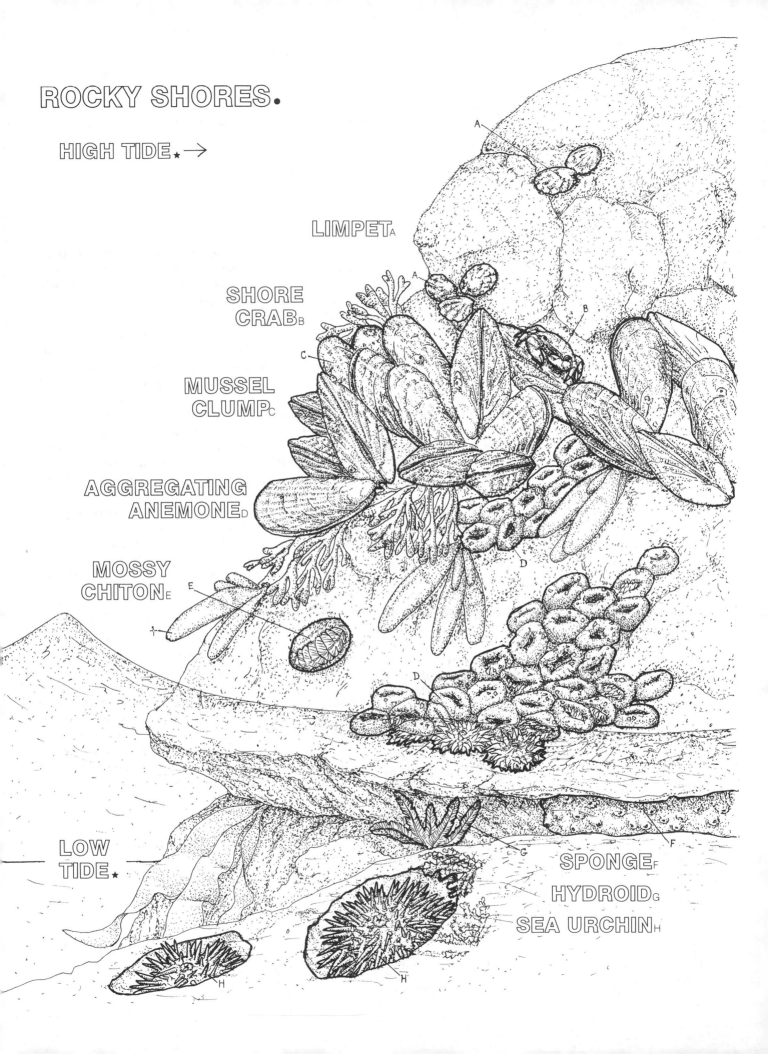

4
CHARACTERIZATION: TIDE POOL

Tide pools are microcosms of the marine life found below the tide line, and are readily accessible when the tide is low. They are formed where depressions in the rock hold reservoirs of sea water. These pools are colonized by sessile (i.e., attached, or stationary) plants and animals, and also by small motile, or moving, forms that either live permanently in the pool or seek their refuge when the tide recedes.

Although these pools are found at every level between the high- and low-tide lines, it is in the pools at the lower levels, where there is less fluctuation in physical conditions, that the most abundant and diverse life forms reside. The organisms discussed here represent the inhabitants of a typical low-intertidal pool on the central California coast.

Color each organism as it is mentioned in the text. You may wish to use the organism's natural color as suggested.

The background coloration in tide pools is created by the substratum and the attached plants and animals residing there. In the well-lighted areas of the pool are the pastel pinks of the *encrusting* and *articulated coralline algae,* the bright green of *surf grass,* and the subtle reds and purples of the delicate *red algae.* In the shadowed areas are the bright reds and yellows of encrusting *sponges* and the vibrant orange of the tiny *solitary corals.* Also tucked in among the shadows is the *giant green sea anemone,* waiting like a deadly flower for unwary prey, or for dead material washed within its grasp.

Moving through this colorful seascape is a multitude of very small crustaceans, molluscs, and worms (not shown). The larger animals include the *sculpin* that darts about briskly and then seems to fade into the background by virtue of its camouflaging mottled coloration. These fishes (5–7.5 cm) venture considerable distances from the tide pool when the tide is in, and often return to the same pool when the tide recedes.

Another well-camouflaged traveler in the pool is the *broken-back shrimp,* capable of rapid movement by the sudden flexure of its powerful abdomen. These crustaceans (2.5 cm) display a range of colors from solid greens to colorful mottled combinations. They are sometimes found in great numbers, but are often difficult to see because of their camouflaging.

The *hermit crab* is fairly easy to spot in the tide pool. This crab occupies an empty sea snail shell, and seeks out a larger shell as it increases in size. If disturbed, the *crab* rapidly disappears into its sanctuary; when unmolested, the *hermit crab* scavenges for food along the bottom of the tide pool.

Bearing a shell of its own creation, the *dunce cap limpet* may be seen feeding on *encrusting coralline algae.* The *dunce cap limpet* is one of the few animals able to consume this unpalatable alga, which is found growing on its own shell.

The small (5–7.5 cm) six-rayed *sea star* may be found patrolling the bottom and sides of the pool. This generalist carnivore feeds on a variety of the small attached life forms in the pool.

If the tide pool has any loose rocks on the bottom where small amounts of sediment have collected, one may find delicate *brittle stars,* and burrows of *polychaete worms.* If sufficient sediment has collected on the bottom, one may also discover *rock crabs,* that have burrowed in, with only their stalked eyes and antennae visible above the sand.

TIDE POOL.

ATTACHED. ★
CORALLINE ALGAEA( )
   ENCRUSTINGA¹
   ARTICULATEDA²
SURF GRASSʙ
SPONGEᴄ
SOLITARY CORALᴅ
GIANT GREEN
   SEA ANEMONEᴇ
RED ALGAEꜰ

MOTILE. ★
HERMIT CRABɢ
TIDEPOOL SCULPINʜ
BRITTLE STARɪ
BROKEN-BACK SHRIMPⱼ
DUNCE CAP LIMPETᴋ
SEA STARʟ
POLYCHAETE WORMᴍ
ROCK CRABɴ

5
CHARACTERIZATION: COASTAL WETLANDS

Along both coastlines of the continental United States are areas that are well protected from the direct onslaught of waves. These areas are designated coastal wetlands and include coastal lagoons, estuaries, and sloughs; the latter two habitats are often supplied with fresh water from rivers or coastal streams. In these quiet waters, soft sediments and organic matter are deposited by rivers and tidal flows, resulting in the development of a soft mud bottom. This plate introduces one of the most conspicuous and important habitats found in coastal wetlands — the salt marsh.

Reserve a light green color for the cord grass (A), blue for tidal channel (B) and brown for mud flat (H). Begin by coloring the single cord grass plant in the center of the page, including the roots. Proceed to color the areas of cord grass in the picture of the salt marsh. Also color the tidal channel and the mud flat. At the top of the page, color the enlargement of the blade of cord grass showing the salt discharge. Color the arrow indicating detritus falling into the tidal channel area. Then color the channel and mud flat and each of its inhabitants as they are mentioned in the text. And finally, color the arrow for animal waste, much of which is recycled and eventually absorbed by the cord grass through its roots.

Salt marshes stretch over millions of acres along the east coast of the U.S. and along the Gulf of Mexico, and they fringe the edges of major estuaries on the west coast. At the heart of the marsh, from the mid-tide to the high-tide lines, is a rugged flowering plant known as *cord grass*. *Cord grass* is one of the few flowering plants that can survive immersion in salt water: excess salt coming into the plant is accumulated and discharged through the leaves.

Cord grass sends its roots deep into the rich marsh *mud,* tapping the rich nutrients and sending out underground stems from which new plants sprout. *Cord grass* can soon monopolize an area; in the Georgia salt marshes, *cord grass* can produce an annual average yield of 80 tons per acre, making it one of the most productive habitats in the world.

Only about 10 percent of the *cord grass* that grows is consumed directly by animals. Most of it dies, breaks apart, and is carried into the *tidal channels* by the ebb and flow of the tides. Here, the plant material is broken down into particles, called *detritus,* and further decomposed by bacteria and fungi. This *detritus* forms the basis of a food chain that literally fuels the major fisheries of the east coast. Much of the *detritus* is transported directly out of the marsh with the ebb tide and finds its way into near-shore food chains.

Among the *cord grass* roots and along the *tidal channels* are *mussels,* which feed by filtering out the *detritus* and minute plants hanging in suspension in the water. *Detritus* is utilized by other filter feeders, such as *oysters* and *clams* inhabiting adjacent *mud flats,* as well as the young *menhaden* fish. The salt marsh forms the hatching ground and "nursery" for the first eight months of the *menhaden's* life, before it goes to sea to join the largest fishery on the east coast.

Small, green *grass shrimp* are abundant in the *tidal channels* and eat the *detritus* found there. These *shrimps,* in turn, fall prey to *flounder* and young *striped bass,* who follow the tide into the salt marsh to feed. The *blue crab* also lurks in the *tidal channels* waiting for prey.

When the tide recedes, the *tidal channels* and the lower *mud flats* are left exposed. Hordes of *fiddler crabs* emerge from their mud *burrows* to sift the rich mud for plant *detritus.* During the mating season, large male crabs stand by the entrances to their *burrows* and wave their outsized claws in an attempt to attract a female partner.

All marine animals in the *channel* and the salt marsh deposit their *wastes* on the marsh bottom, where they are then decomposed by bacteria and recycled into the *cord grass* through its roots.

All around the salt marsh the sounds of shorebirds and migrating ducks and geese may be heard. The salt marsh serves a vital role in linking the land with the sea. The marsh's productivity and its role as a nursery area figure importantly in the success or failure of near-shore fisheries.

COASTAL WETLANDS.

SALT MARSH ★

CORD GRASS A
TIDAL CHANNEL B
MENHADEN C
GRASS SHRIMP D
STRIPED BASS E
BLUE CRAB F
FLOUNDER G

MUD FLAT H
OYSTER I
CLAM J
MUSSEL K
FIDDLER CRAB L
BURROW M ★

DETRITUS A¹
ANIMAL WASTE B¹

SALT DISCHARGE ⊹

Sandy beaches, a familiar sight along open coastlines, form by the accumulation of *sand* particles, the product of erosion, which have been carried and deposited by *waves.* Moving water carries particles suspended in it, and the more rapid the water movement, the larger the particles carried. When the water slows down, the largest and most dense particles are the first to settle out. As the water becomes very still, the tiniest particles settle out (sedimentation or deposition) and accumulate as mud or silt.

On the open coast, where the water is seldom completely quiet, the larger particles settle out to form sand bottoms. A typical mainland beach is composed of small particles and fine gravel of quartz and feldspar. The beaches of tropical islands are sometimes composed of eroded coral, and the black sands of certain Hawaiian beaches came from the erosion of lava flows. The particles that form these sandy shores originate on the land surface and are carried into the sea by rivers, by wind, or are weathered from the shoreline by the pounding *waves.*

Once a beach forms, it changes continuously; through the seasons, the *waves* constantly rework the *sand* and reshape the beach.

Color the illustrations of the sandy beach, shown in spring, summer, fall, and winter. The curved arrows in the fall illustration represent the wave-driven removal of sand from the berm.

During spring and summer, gentle *waves* deposit *sand* onto the *beach platform,* forming a broad sandy slope or *berm.* The large *waves* of the first fall storm begin to remove *sand* from the beach and deposit it offshore in ridges, called sandbars. The winter beach may have nothing remaining but the rocky *beach platform* (the eroded edge of the coastline) and *cobblestones* too large for the winter *waves* to carry off.

The beach is obviously a rigorous environment, and organisms that live here must adapt to constantly shifting *sand* and *waves.* Successful species are able to ride the *waves* and to burrow deep into the *sand,* or are able to reside just above the tide line. Sandy shores support relatively few species, which, however, generally occur in great numbers due to reduced competition between species. The *waves* bring a steady source of fine organic matter (detritus) as well as such larger pieces of organic debris, as loose seaweed and dead fish. These materials provide a dependable food source for the marine organisms of the sandy shore.

Color each sandy-beach organism as it is mentioned in the text, coloring both the large and small illustrations of the animals.

The *sand crab* uses the waves in moving up and down the shore. It feeds by burrowing its posterior end into the sand and then unfurling its long antennae into the overlying water to filter food from the wave backwash (see Plate 29).

The small (2.5 cm) wedge-shaped *bean clam* of east coast beaches is a rapid burrower and rides the waves, staying in the area of greatest water movement, where suspended food is most abundant.

The *razor clam* is found along the beaches of the Pacific Northwest near the low-tide line. *Razor clams* thrive on surf-zone diatoms (microscopic plants) that bloom throughout spring and summer. Anyone who has pursued these delectable clams can tell you of their burrowing speed.

Like the *razor clam,* the *bristle worm,* a scavenging polychaete, remains in the lower zone of the beach and relies on its rapid burrowing to keep from being washed away by the tide.

Beach hoppers live in burrows above the high-tide line and emerge at night to feed on deposited drift algae. The predaceous *rove beetle* also lives above the high-tide line. It comes down onto the shore at night in pursuit of the unwary *beach hopper.*

Another nocturnal carnivore is the *ghost crab* of tropical and semitropical sand beaches. These *crabs* are highly tolerant to air exposure and live in burrows above the high-tide line.

The *swimming crab* and the barred *surfperch* ride in with the tide to feed on *sand crabs* and other burrowing animals.

SANDY BEACH.

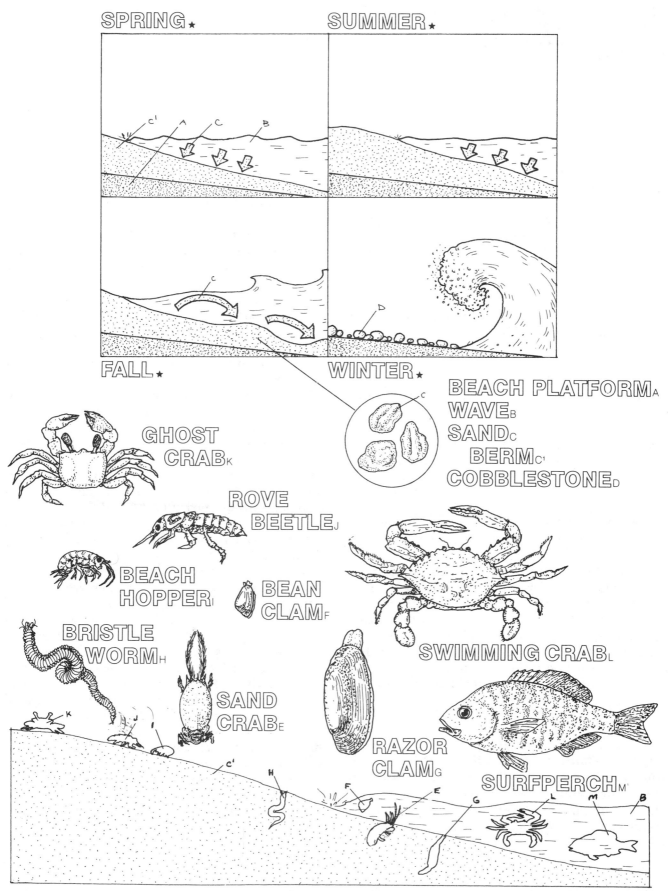

SPRING★ SUMMER★

FALL★ WINTER★

BEACH PLATFORM_A
WAVE_B
SAND_C
BERM_{C1}
COBBLESTONE_D

GHOST CRAB_K

ROVE BEETLE_J

BEACH HOPPER_I

BEAN CLAM_F

SWIMMING CRAB_L

BRISTLE WORM_H

SAND CRAB_E

RAZOR CLAM_G

SURFPERCH_M

7
CHARACTERIZATION: SUBTIDAL SOFT BOTTOMS

Most of the ocean floor consists of soft sands and muds. The distribution of soft bottom types is generally controlled by the same factors that create the sandy beaches: water movement and the character of the materials held in suspension determine the type and composition of the bottom.

Soft bottoms on the continental shelf are composed mainly of inorganic materials washed from the land by rivers, or carried away by wind. The shallow bottom closest to shore experiences the strongest wave action and water currents; these keep all but large and dense sediment particles in suspension. These particles settle out, forming near-shore sandy bottoms. These same turbulent near-shore waters hold a large amount of organic detritus in suspension. The faunas of these sand bottoms are dominated by filter feeders and their predators.

At greater depth or in sheltered areas, water movement is lessened, and the finer suspended particles are deposited. A gradient of soft bottoms results, ranging from coarse sand near shore, to muds, composed of very fine silt and clay particles found offshore over the continental shelf. As the water becomes calmer and the suspended organic detritus settles to the bottom, the fauna shifts from suspension, or filter feeders, to deposit feeders, who feed on this material.

Beyond the continental shelf, the character of the soft bottom reflects what lives and grows above it in the water column. Vast areas of deep ocean bottom consist of the skeletons of minute planktonic plants and animals (diatoms, foraminiferans, and so on), which form muddy "oozes" many feet thick. Other areas of the deep ocean bottom are covered with red muds composed of clay particles that have settled out very slowly.

Color each of the animals as it is mentioned in the text. Color both the illustrations of the animal in its environmental position (at the top of the page) and the larger illustrations below.

The soft bottom does not offer suitable substratum for the attachment of large algae (seaweed), and the main plants living here are specialized benthic (bottom dwelling) diatoms that live on the surface of the mud. The animals here are divided into an epifauna, living on the surface, and the infauna, which live in the substratum.

Residing just below the low-tide line along the central California coast is the large (15 cm) *Pismo clam*. *Pismos* live in clam beds and feed on the large amount of suspended detritus in the surf zone. They depend on their heavy shells to keep them in place.

Just beyond the surf zone, beds of the suspension-feeding epifaunal *Pacific sand dollar* may be found, together with the scavenging *elbow crab* and *hermit crab*. Predators such as the *sand star* and the *moon snail* occur here, feeding on the numerous species of filter-feeding clams, like the *sea cockle*. Other life forms include the Pacific *sanddab* flatfish and the flat *angel shark*.

In deeper, calmer waters, large aggregations of *brittle stars* may be found swarming on the bottom, feeding on deposited material, or buried in the sediment, with only their arms protruding through the mud. Another spiny-skinned animal found here is the burrowing *heart urchin,* or sea porcupine, which ingests sediment as it burrows, digesting the organic material contained therein.

Worms are among the most abundant infaunal animals. Their sleek, elongated bodies are perfectly adapted for efficient burrowing in soft substrata. Many types of worms are found within the soft substrata; some ingest the sediment as they burrow through it; others feed on detrital materal deposited on the surface; some filter suspended materials from the water above. Shown here is a robust (5–10.0 cm) burrowing *polychaete worm,* a carnivorous scavenger that re-burrows very rapidly if uncovered.

Soft substrata are very monotonous, homogeneous habitats that offer a limited number of kinds of places for plants and animals to live, and thus, these habitats house a limited number of kinds of organisms. However, individual members of the fauna can be extremely abundant, and play important roles in marine food webs.

SUBTIDAL SOFT BOTTOMS.

EPIFAUNA ★
PACIFIC SAND DOLLAR A
SAND STAR B
MOON SNAIL C
ELBOW CRAB D
SANDDAB E
ANGEL SHARK F
BRITTLE STAR G
HERMIT CRAB H

INFAUNA ★
PISMO CLAM I
SEA COCKLE J
HEART URCHIN K
POLYCHAETE WORM L
BRITTLE STAR G

8
CHARACTERIZATION: KELP BED

The kelp bed is a most productive and interesting cold-water marine habitat. Kelp beds are near-shore areas dominated by the presence of very large brown algae, called kelps. These plants require cold sea water and a solid rocky bottom on which to attach. They also require significant water movement to insure a constant supply of dissolved nutrients is available to fuel their photosynthetic processes. Kelp beds commonly grow in about 20 meters of water; if the water is exceptionally clear, they may grow at depths up to 30 meters. The kelp bed structure and some of its more conspicuous inhabitants are discussed here.

Color each organism as it is introduced in the text. Note that the invertebrates and fishes in the foreground are drawn on an exaggerated scale relative to the large kelps for purposes of illustration. The sea lion and sea otter are drawn in the background to suggest the depth of the kelp bed.

The kelp bed is very much like an underwater forest. Off the coast of southern California, the dominant kelp is the *giant kelp*. Stipes of the *giant kelp* can grow to lengths of 30 meters from its steadfast anchor on the rocks to the surface of the water. Its blades spread to form a thick canopy that soaks up the sun's energy for photosynthesis. These giant plants are kept afloat by the bulb-shaped air bladders (pneumatocysts) at the base of each blade.

Beneath the overhead canopy created by the *giant kelp*, the smaller *palm kelp* grows. This species has a thick elastic stipe that bends with the water move-

ment. The *palm kelp* uses the light that filters through the overhead canopy for its photosynthesis.

On the rocky bottom of the kelp bed is a turf-like layer of small *red algae*. If the area is very densely shaded, it may contain a variety of attached invertebrates: *sponges,* sea *anemones,* sea squirts, and barnacles (not shown). Among these attached creatures live millions of smaller, motile animals. Brittle stars, gastropods, amphipods, and isopods abound.

The herbivorous (plant-eating) sea *urchins, sea hares,* and *abalones* can often be found on the bottom, where they take advantage of the large amounts of plant material produced in the kelp bed. The omnivorous (plant and animal-eating) *sea bat* is also a conspicuous kelp bed resident, as is the large carnivorous *sunflower star,* which eats the prickly sea *urchin,* other sea stars, and a variety of other invertebrates.

The waters below the kelp canopy are rich in fish life. The boldly colored *sheephead* comes to feed on larger invertebrates that live among the kelp stipes. The kelp bed is also home to several species of *rockfishes,* which feed on other fishes in the bed and a variety of invertebrates. Individual *rockfish* species occupy relatively discrete sub-habitats within the kelp bed, thus avoiding direct competition with each other.

Two marine mammals that frequent kelp beds are the *sea lion* and the *sea otter*. The *sea lion* stalks fish and sometimes plays between the large kelp stipes. The *sea otter* may spend almost its entire life in the kelp bed and plays a very important role in this habitat (as will be discussed in Plate 96.)

KELP BED.

SEA OTTER_N

GIANT KELP_A

SEA LION_M

SHEEPHEAD_K

ROCKFISH_L

PALM KELP_B

RED ALGAE_C

SEA HARE_G

URCHIN_F

SEA BAT_I

ANEMONE_E

SPONGE_D

ABALONE_H

SUNFLOWER STAR_J

9
CHARACTERIZATION: FRINGING REEF

The coral reef is one of the most complex of under-water seascapes. Warm, clear water, ranging from 20° to 23°C, is required for coral reef formation. This condition is found only in tropical and semitropical waters between 30° North and 30° South latitudes of the Indian Ocean, the tropical Pacific, and the Caribbean Sea. In these areas, three main types of coral reef may form: the fringing reef, the barrier reef, and the atoll.

Our discussion here centers on the most common type of reef, the fringing reef. This reef extends seaward from shore, and is found surrounding islands and bordering continents. The fringing reef displays a zonation pattern corresponding to the depth and the turbulence of the water. The illustration represents a wide range of depth.

Begin by coloring in the upper picture as each organism is mentioned in the text. The upper picture represents the animals found on the coral reef in the daytime; the bottom represents the same setting at night.

Coral reefs are constructed by corals, which are primitive animals belonging to the coelenterate group that also includes jellyfishes and sea anemones. The individual coral animal, or polyp, looks very much like a sea anemone and feeds on zooplankton in the water. Coral polyps exist in a mutually beneficial relationship with single-celled plants called zooxanthellae (see Plate 76). The zooxanthellae live within the coral tissue and facilitate the coral polyp's secretion of a calcium carbonate (lime) skeleton. It is this skeleton that forms the basis of the coral reef. The zooxanthellae also provide the coral polyp with nutrients to supplement its diet of zooplankton, while the coral supplies the alga with certain essentials for photosynthesis, as well as a place to live. Only the surface of the reef harbors live coral polyps. The layers beneath are the skeletal remains of dead corals, which have become compacted and fused together.

In the uppermost level of the reef, the large *elkhorn coral* predominates. Each coral formation is an entire coral colony that may be many meters across, the pro-duct of hundreds of years of coral growth. Coral branches often grow in the direction of the prevailing water current.

Deeper beneath the water surface, massive corals like the mound-forming *star coral* and the *brain coral* are found. In areas of decreased light penetration, the *plate coral* grows, spreading out flat to maximize exposure to remaining light, which is needed by the zooxanthellae for photosynthesis.

The reef supports a great many species of corals, each competing for attachment space and maximum light exposure. This tangle of growth results in a maze of crevices, caves, and ledges, which harbor other marine animals, including *sponges* and soft corals like the *sea fans*.

The coral reef supports both daytime and nightime inhabitants. During the day, the polyps of the *brain coral* and *star coral* are retracted, but at night they extend to trap small zooplankton. During the day, *groupers, butterflyfish, damselfish, parrotfish,* and others swim busily about the reef. *Cleaner shrimps* are found waving their long antennae, waiting for the fishes to visit their "cleaning stations."

As night approaches, the daytime feeders take refuge in holes or crevices in the reef; the *parrotfish* secretes a protective thin mucous coating around its body. Out from their daytime refuge come the big-eyed *squirrelfishes* and the brightly striped *grunts* to begin their nocturnal foraging. The *moray eels* also come out to feed.

Some large invertebrates also emerge at night: the *feather stars* climb to a suitable perch and unfurl their many arms to begin filter feeding. The *spiny lobster* crawls out from under a deep crevice or overhang and scavenges about the reef for food. The long-spined *sea urchin,* visible by day only as a phalanx of formidable spines protruding from a crevice, moves with surprising speed towards the small algae that it feeds on.

As daylight appears, the nocturnal animals return to their daytime retreats; the coral polyps once again retract; and the coral reef recommences its daytime activity.

FRINGING REEF.

ELKHORN CORAL_A
STAR CORAL_B
BRAIN CORAL_C
PLATE CORAL_D
SPONGE_E
SEA FAN_F

DAY SHIFT:
GROUPER_G
BUTTERFLYFISH_H
DAMSELFISH_I
PARROTFISH_J
CLEANER
    SHRIMP_K

NIGHT SHIFT.
SQUIRRELFISH_L
GRUNT_M
MORAY EEL_N
FEATHER STAR_O
SPINY LOBSTER_P
SEA URCHIN_Q

10
CHARACTERIZATION: PELAGIC ZONE

The pelagic zone is the most extensive of the marine environments; it encompasses the volume of water from just above the ocean floor to the surface, and it extends from shore to mid-ocean. The pelagic environment is divided according to its proximity to the continents and depth of water. This plate is concerned with the upper layer of the pelagic zone, where light penetrates through the water.

Light penetration is necessary for the growth of plants and for the presence of animals that must see in order to feed. The water temperature has a great influence on organisms' growth rate and metabolism, and water movement plays an important role in the pelagic zone where organisms either drift with the water current or swim within it.

Begin by coloring the three drifting animals belonging to the neuston group as they are discussed in the text.

Neuston are those organisms that float on the surface of the water. Although *neuston* are moved by both *wind* and surface *currents,* the *wind* is the most important mover of these animals. One of the most common coastal water drifters is called the *by-the-wind-sailor.* The *by-the-wind-sailor* (a coelenterate) is actually a whole colony of individual organisms that live, feed, and float together. As a colony, they set *sail* to the *wind,* producing a thin blue membrane of stiff material, oriented diagonally across the colony's elliptical body. As *wind* carries the colony across the surface, one individual in the center of the colony feeds, while others function in defense or reproduction.

Another colonial coelenterate is the *Portuguese man-of-war,* which utilizes a *gas*-filled *sac* as both a float and a sail, while long prehensile tentacles dangle into the water to catch prey.

The *violet snail* is also a member of the *neuston* group. The *snail* secretes *bubbles* that keep it afloat. This animal feeds on the *by-the-wind-sailor.*

Now color the enlarged plankton which are moved by the current.

Plankton includes drifting organisms that are found at most depths in the water column. The single-celled *phytoplankton* are minute plants, and are successful only in the uppermost section of the pelagic zone where there is plenty of light for photosynthesis. Zooplankton (animal plankton) species, such as *copepods* and *euphausiids,* feed on the abundant *phytoplankton.* The predatory *arrow worm,* or chaetognath, is known as the "tiger of the zooplankton" for its voracious appetite. These small animals dart about and devour prey as large or larger than themselves. Zooplankton play an important role in the food chain and are the main source of nourishment for the young of many fish species, as well as much larger animals that strain the water for food.

Color the nekton which are self-propelled.

Nekton includes those pelagic dwellers that can swim in directions independent of the ocean's currents. This category includes the earth's largest mammal, the *blue whale* (27 meters in length). The *blue whale,* together with several other whale species, is a filter feeder, whose principal food source is the zooplankton, especially *euphausiids,* or "krill." The *nekton* include many fast and efficient swimmers, such as the *squids,* which utilize their own variety of jet propulsion. The more conventional *albacore* prey on the smaller *herring.* The sharks are a highly successful and predatory group, represented here by the *blue shark.*

The *phytoplankton* growing in the sunlit upper layers of the water column provide the basis of virtually all life in the pelagic zone. The productivity of the *phytoplankton* thus governs the number of organisms that can survive in the pelagic zone. This explains why, with increasing depth or distance from land and its input of nutrients, proportionally few creatures are found.

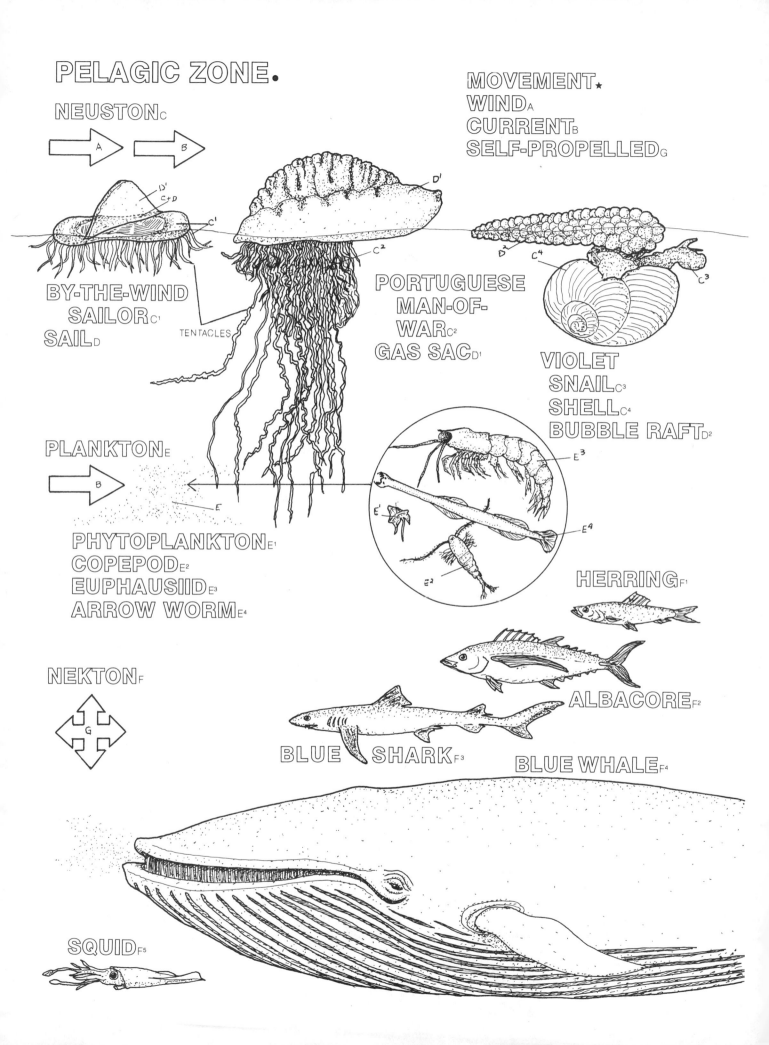

PELAGIC ZONE.

NEUSTONc

A → B →

MOVEMENT★
WINDA
CURRENTB
SELF-PROPELLEDG

BY-THE-WIND SAILORC1
SAILD

TENTACLES

PORTUGUESE MAN-OF-WARC2
GAS SACD1

VIOLET SNAILC3
SHELLC4
BUBBLE RAFTD2

PLANKTONE

B →

PHYTOPLANKTONE1
COPEPODE2
EUPHAUSIIDE3
ARROW WORME4

HERRINGF1

NEKTONF

G

ALBACOREF2

BLUE SHARKF3

BLUE WHALEF4

SQUIDF5

11
MARINE FLOWERING PLANTS

Most plants become metabolically stressed if exposed to sea water, as they are incapable of dealing with the high salt content. This metabolic stress ultimately results in the death of the plant.

A few species of land plants have, however, successfully invaded and adapted to sea-water environments, and flourish in a variety of marine habitats. Unlike algae, land plants can grow only where they can absorb nutrients through their roots. These higher plants also need nearly direct sunlight, so they can only tolerate relatively shallow waters (1–30 meters).

Color the red mangrove, starting with the prop roots at the lower right. Notice the junction of the trunk and roots at the high-tide mark. Color the entire leaf mass green. Color the falling seed and new root protruding from it, and notice the four young leaves emerging from the seed as you color.

One of the largest flowering plants to flourish in the marine environment is the red mangrove. These plants occur in tropical and semitropical regions from Florida to South America and reach tree size (1–3 + meters). The mangrove grows best in the mud bottoms of estuaries, coastal lagoons, and near the mouths of large rivers, where silt is deposited. The mangrove remains stationary in the unstable mud bottom by sinking a mass of large, arching *prop roots* deep into the mud. The *prop roots* divide off from the *trunk* at the high-tide level; the *roots* proliferate and trap more sediment, raising the substratum level. This, in turn, creates more space where the mangrove can grow, and, in this way, the mangrove colony expands toward the sea.

The mangrove *roots* are habitat for a myriad of attached algae and animals such as sea squirts, sponges, and sea anemones. The lattice-work of *roots* also offers protected nursery areas to young reef fishes and spiny lobsters. The mangrove *leaves* fall into the water and decompose, providing a food source for many animals residing near the mangrove, and others further out to sea.

Seeds of the red mangrove sprout and begin to grow before they fall from the tree. A long (36 cm), slender *root* emerges from the *seed,* and several *leaves* may also grow. If the *seed* falls at low tide, it may poke, dart-like, into the soft mud near the parent plant, and continue to grow. *Seeds* that fall during high tide float upright in the water and may be carried miles from the parent plant. When the tide recedes, if the *seed* has been left on the mudflat with the *root* facing downward, it will be "drilled" into the mud by the jostling of small waves, thus beginning a new colony of mangroves.

Now color the turtle grass and the surf grass. Note that the grasses' stems receive the same color as the mangrove's trunk, and the blades receive the same color as the mangrove's leaves. These structures are homologous (same origin) and have the same function in these higher plants.

The flowering plants known as sea grasses can grow completely submerged. Various sea grass species occur from the intertidal zone to depths of 30 meters. Turtle grass is found in eastern Florida, along the Gulf of Mexico, and in the Caribbean, where it thrives on silt bottoms of mud, shell, or sand. Once established, turtle grass proliferates by extending underground *stems* (rhizomes) from which leafy *blades* grow. Turtle grass forms extensive shallow-water meadows, which are frequented by numerous juvenile reef fishes. At night, parrotfish and sea turtles come into these meadows to feed. The *stems* and *roots* of turtle grass bind the substratum, which permits the colonization of burrowing animals, such as sipunculid worms and sea cucumbers (see Plates 18 and 33).

Unlike most sea grasses, which require quiet waters, surf grass grows in the wave-swept rocky intertidal zone of temperate waters. Surf grass *seeds* possess two stiff, bristled, pointed projections. If the *seed* lodges on an erect coralline alga, it can germinate and send out tenacious *roots* to grab hold of the substratum. The surf grass *roots* trap sediment, and a patch of long, thin, bright green *blades* may develop in a tide pool or at the low-tide line. This creates an environment that shelters many species of worms and other small creatures.

MARINE FLOWERING PLANTS.

RED MANGROVE ★

PROP ROOT_A
TRUNK_B
LEAF_C
SEED_D

TURTLE GRASS ★

ROOT$_{A^1}$
STEM$_{B^1}$
BLADE$_{C^1}$

SURF GRASS ★

ROOT$_{A^2}$
STEM$_{B^2}$
BLADE$_{C^2}$
SEED$_{D^1}$

ALGA

HIGH TIDE ★

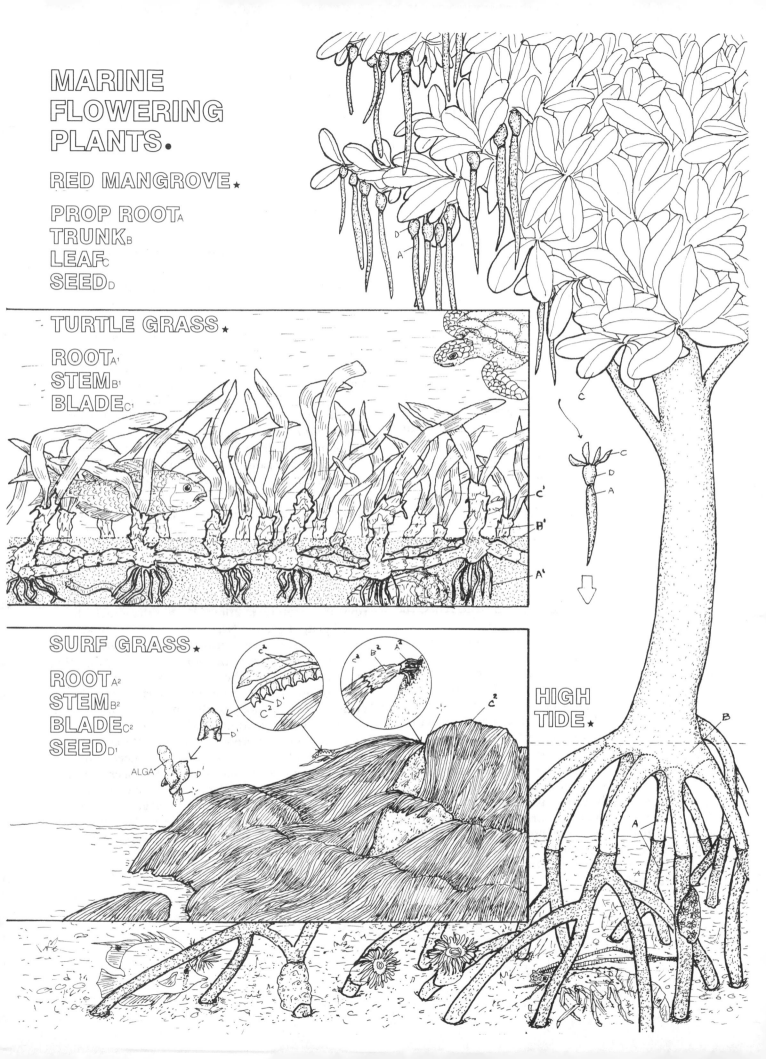

12
PHYTOPLANKTON: DIVERSITY AND STRUCTURE

Beyond the shallow depths where marine flowering plants and the macroalgae grow, the sunlit layer of water is dominated by single-celled plants known as phytoplankton. Phytoplankton includes a variety of plant forms, all of which are autotrophic: they capture energy from sunlight, and require nutrients (phosphates, nitrates, etc.) and carbon dioxide (CO_2) for photosynthesis. A single liter of rich coastal sea water may contain dozens of different species of phytoplankton, and possibly as many as 10–20 million individual one-celled plants. Some species are small, flagellated forms, much too tiny to be captured in any animal's finest-mesh plankton net. The larger, more common phytoplankton species are diatoms and dinoflagellates, which are found abundantly in temperate waters. These species can grow up to 1 millimeter across, but most are much smaller.

Begin by coloring the diatoms in the upper left. Color each diatom structure as it is discussed in the text, using a dark color for the pores (C).

Diatoms are found in both marine and freshwater habitats. Marine diatoms are of two basic types: the elongated forms (Pennales or pennate diatoms), such as *Pleurosigma*, which are usually found in very shallow areas; and the round or wheel-shaped forms, such as *Coscinodiscus* (Centrales or centric diatoms). In the case of *Coscinodiscus,* you see that the diatom consists of a two-part *frustule* which is made of silica and appears like a glass jewel when viewed under the microscope. On the top of the *frustule,* an elaborate pattern of *pores* radiates out from the center; the *pores* help reduce the weight of the floating diatom and allow diffusion of materials into and out of the cell. Viewed from the side, the shape of the *frustule* can be seen: the upper half, or *epitheca,* fits over the smaller bottom half, the *hypotheca.* Inside the *frustule* are the *nucleus* of the diatom, which contains the genetic material, and the *chloroplasts,* or photosynthetic organelles.

Now, color the dinoflagellates in the upper right corner.

Unlike the diatoms, the dinoflagellates have long *flagella* that are used for locomotion. The whip-like *flagella* are located in grooves — the longitudinal *sulcus* and the transverse *cingulum* — on the dinoflagellate. Dinoflagellates have a multi-layered covering of cell material. In the armored (thecate) dinoflagellates, such as *Peridinium,* the cell is encased in an expandable, overlapping layer of cellulose *plates*; this is absent in naked, or unarmored dinoflagellates, such as *Gymnodinium.* Many dinoflagellates are known to be bioluminescent (light-producing), and this group also includes the organisms that cause the sea water to turn red during the so-called "red tides" (Plate 58).

Next, color the various species of the diatom genus *Chaetocerus*. Note the difference in the length and shape of the setae.

The dinoflagellates are able to swim and move up and down in the water column; diatoms cannot move under their own power, but have developed adaptations that keep them afloat. Within the widely distributed diatom genus *Chaetocerus,* a variety of adaptations are visible. The individual *Chaetocerus* cell is oval, with two pairs of thin spines or *setae* projecting from either end of the cell. These *setae* fuse with those of other cells to form long chains, thereby increasing the buoyancy of the chained group. As you see from the illustration, the length and shape of the *setae* vary with different species. *Chaetocerus decipiens* is found in cool, dense water, and needs only relatively short *setae* to stay afloat. *Chaetocerus* species living in warmer, less dense water have developed long *setae* that provide more resistance to sinking. *Chaetocerus denticulatus* has secondary spines on the *setae* to keep it from sinking in warm water.

PHYTOPLANKTON: DIVERSITY & STRUCTURE.

DIATOMS★
PLEUROSIGMA A

COSCINODISCUS B( )
FRUSTULE B¹
PORES C
EPITHECA D
HYPOTHECA E
PROTOPLAST★
NUCLEUS F
CHLOROPLAST G

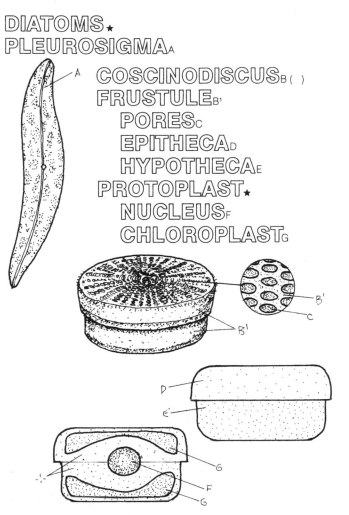

DINOFLAGELLATES★
PERIDINIUM J( )
FLAGELLUM K
SULCUS L
CINGULUM M
THECAL PLATES J¹

GYMNODINIUM N

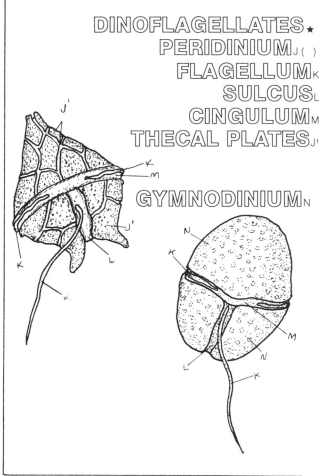

SPECIES OF CHAETOCERUS H( )
SETA I

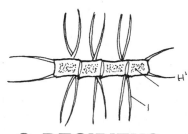

C. DECIPIENS H¹

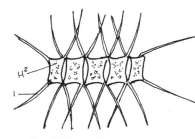

C. DIDYMUS H²

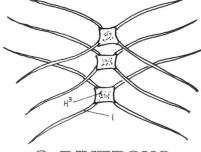

C. DIVERSUS H³

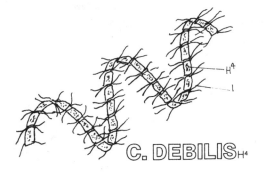

C. DEBILIS H⁴

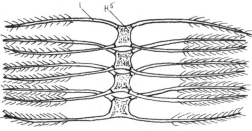

C. DENTICULATUS H⁵

13

SEAWEED ADAPTATIONS: RED AND GREEN ALGAE

The multicellular algae, or macroalgae, have evolved some very important adaptations allowing them to survive in the near-shore shallow marine environments. Macroalgae are limited to areas of shallow water and rocky shores; the algae need light for photosynthesis, and solid substrata for attachment sites. But the near-shore habitat poses a challenge to the algae; wave action, desiccation during low tide, and grazing by numerous herbivores are all potential threats to these plants.

Begin by coloring the sea lettuce. The illustration at the upper right shows various algae attached to a rocky surface; the other illustrations are larger drawings of the algae. As each type is mentioned in the text, color both representations. Choose greens for the green algae, pink for the coralline algae, and reds and purples for the red algae.

The green algae are abundant in freshwater environments, but are represented by relatively few marine species. One green alga of the marine intertidal zone is the thin-bladed (two cells thick) *sea lettuce*. The delicate *sea lettuce* loses its moisture during low tide and becomes quite crispy, yet remains alive. When the tide rises, it again absorbs water. The green alga *Cladophora* grows in small, thick tufts in the middle intertidal zone. These tufts are composed of thousands of tiny multi-branched filaments that serve to trap sand and to hold precious water for survival during low-tide exposure.

The red alga known as the *sea sac* is a medium-sized (2.5–5 cm in diameter) alga of the middle intertidal zones, often found in large patches. The *sea sac* is aptly named, for in its hollow core it holds a reservoir of sea water to keep it from drying out at low tide. These water sacs are home to a particular type of tiny copepod (crustacean).

The grazing activities of invertebrate herbivores can severely injure an alga, and several of the alga species have evolved means by which they combat grazing. The *coralline red algae* secrete a calcium carbonate (lime) in their cell walls, making them a tough, crusty plant. Most herbivores avoid eating the *coralline red algae,* with the notable exception of the dunce cap limpet.

Two general types of *coralline red algae* are present in the marine intertidal zone. The *encrusting corallines* occur in shady areas of tide pools and cover the rocks in bumpy-textured pink sheets. The jointed or *articulated coralline algae* also occur in the lower intertidal zone. They are small (5–7.5 cm), erect, branched plants that have calcified sections interlaced with flexible joints. The overall calcified nature of the *articulated corallines* offers resistance to the pounding waves, with enough flexibility to bend with the water movement.

Many red algae are highly branched, with beautiful and intricate patterns of growth. This growth pattern serves to increase the light-gathering surface of the plant for photosynthesis, and also makes it more difficult for small grazers to attach themselves to the branched structure. One such alga is the *pepper dulce.* As its common name implies, the plant has a biting, peppery taste; the plant may possibly sequester noxious chemicals in its tissues, making it still more unappealing to herbivores.

The intertidal area is often crowded with plant life, and there are few open spaces to which an alga can attach. Some algae then attach to other plants or animals. One prominent example is the red alga *Smithora,* which is found only attached to eel grass and to surf grass. *Smithora* has a small (1–2 cm) reddish-purple blade that may entirely cover individual surf grass blades (as shown). Biologists believe that this arrangement may be mutually beneficial — grazing herbivores prefer to feed on the fleshy, more prominent algae and leave the surf grass alone.

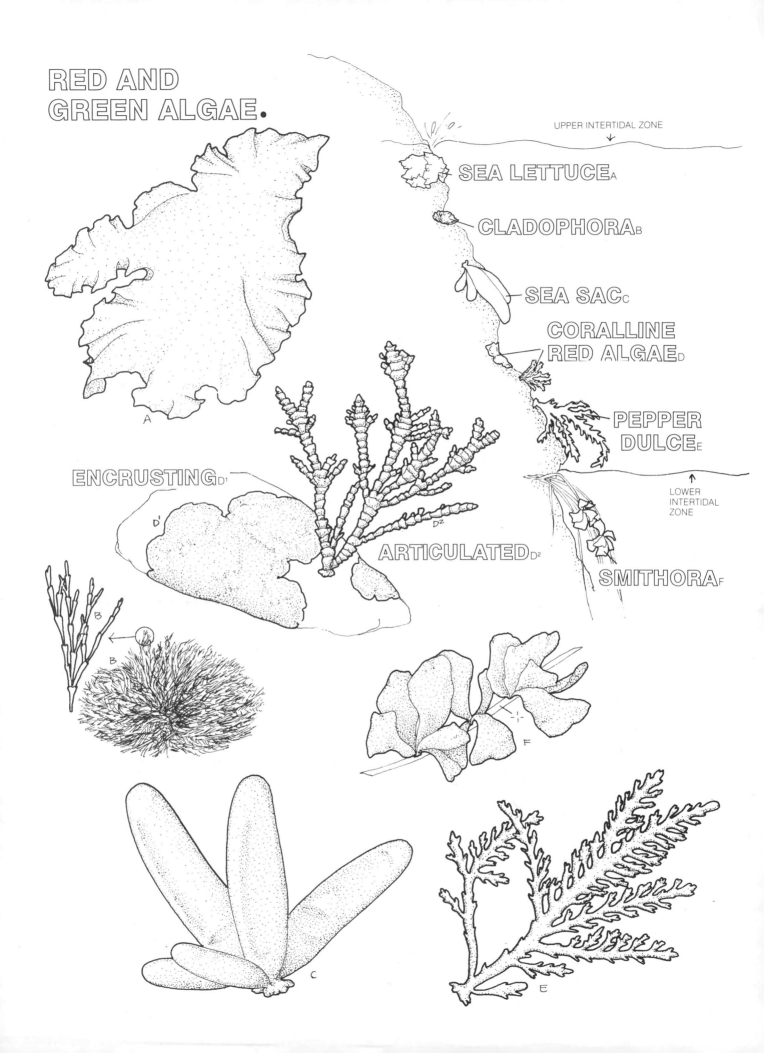

RED AND GREEN ALGAE.

UPPER INTERTIDAL ZONE

SEA LETTUCE A

CLADOPHORA B

SEA SAC C

CORALLINE RED ALGAE D

PEPPER DULCE E

ENCRUSTING D¹

ARTICULATED D²

SMITHORA F

LOWER INTERTIDAL ZONE

14
SEAWEED ADAPTATIONS: BROWN ALGAE

The brown algae are subjected to the same environmental stresses that exist for the red and green algae: damage from wave shock, desiccation, grazing by herbivores, competition for available attachment space and for light for photosynthesis. Some brown algae are much larger than other types of algae; this is often a survival advantage, and brown algae have developed several other mechanisms as well.

Begin by coloring the rockweed at the upper left, coloring each part of the alga. Color each brown alga as it is treated in the text. Notice the differences in relative proportions of the holdfast, stipe, and blade among these plants. Note that the feather-boa kelp has especially long stipes and small blades.

Rockweed is a common brown alga worldwide, growing in the high and middle zones of the rocky intertidal. These zones are a potentially very stressful area, being most frequently exposed to air at low tide. The rockweed is successful here because it can tolerate considerable desiccation (drying out), such an event being retarded by thick cell walls and high concentrations of polysaccharides (sugars). Rockweed does not grow structures that are readily distinguishable as *stipes* or *blades*; the entire plant is called a *thallus*. At high tide, small air bladders (not shown) along the sides of the *thallus* cause the alga to float up from the bottom where there is better exposure to light. The swollen tips of the *thallus* are reproductive structures called *receptacles*.

The large brown algae known as kelp are common subtidally and in the low intertidal zone, where the wave action is most severe. The *feather-boa kelp* has long (10 m), supple, strap-like *stipes* that wash back and forth with the movement of the water. The *stipes*

do sometimes snap off during storms, leaving the large *holdfast* behind. The *holdfast* resembles a mound of intertwined roots and serves to anchor the kelp. It can persist over a winter, and in the spring, new *stipes* will sprout.

Another wave-zone kelp is the oar weed, a type of *Laminaria* (see Plate 2). It, too, has a stout *holdfast,* and limber, hollow *stipes*. The resiliency of the *stipes* serves to keep its deeply incised *blades* upright for better exposure to sunlight.

Lessoniopsis is a kelp found in the most exposed, wave-battered areas of the low intertidal. Its huge *holdfast,* resembling a thick woody trunk, enables it to survive in this environment. *Lessoniopsis* is a perennial alga that may persist for many years. Each year it adds another layer to its "trunk," which can grow to 20 centimeters thick. The *stipe* of *Lessoniopsis* is highly branched and sports long (1 m), thin *blades*. A large plant may have over five hundred *blades*.

The bull kelp is one of the "giant kelps." It is an annual alga (lives for a single year), and occupies the shallow subtidal zone to depths of 30 meters. The bull kelp possesses an elongated, rapidly growing *stipe* (10 centimeters a day under ideal conditions) which ends in a large, gas-filled *air bladder* (pneumatocyst), which serves to lift the plant to the water's surface. Large *blades* grow from the pneumatocyst and spread out and take advantage of the sunlit water. In the fall and winter, storm waves dislodge the bull kelps and wash them onto the shore in a tangled mass.

Humans have put the rapidly growing giant kelps to use. In southern California, large barges with paddle-wheel mowers harvest the kelp beds. The kelp is rendered to yield algin — a stabilizer and emulsifier used in many products, including paint, ice cream, and cosmetics.

BROWN ALGAE.

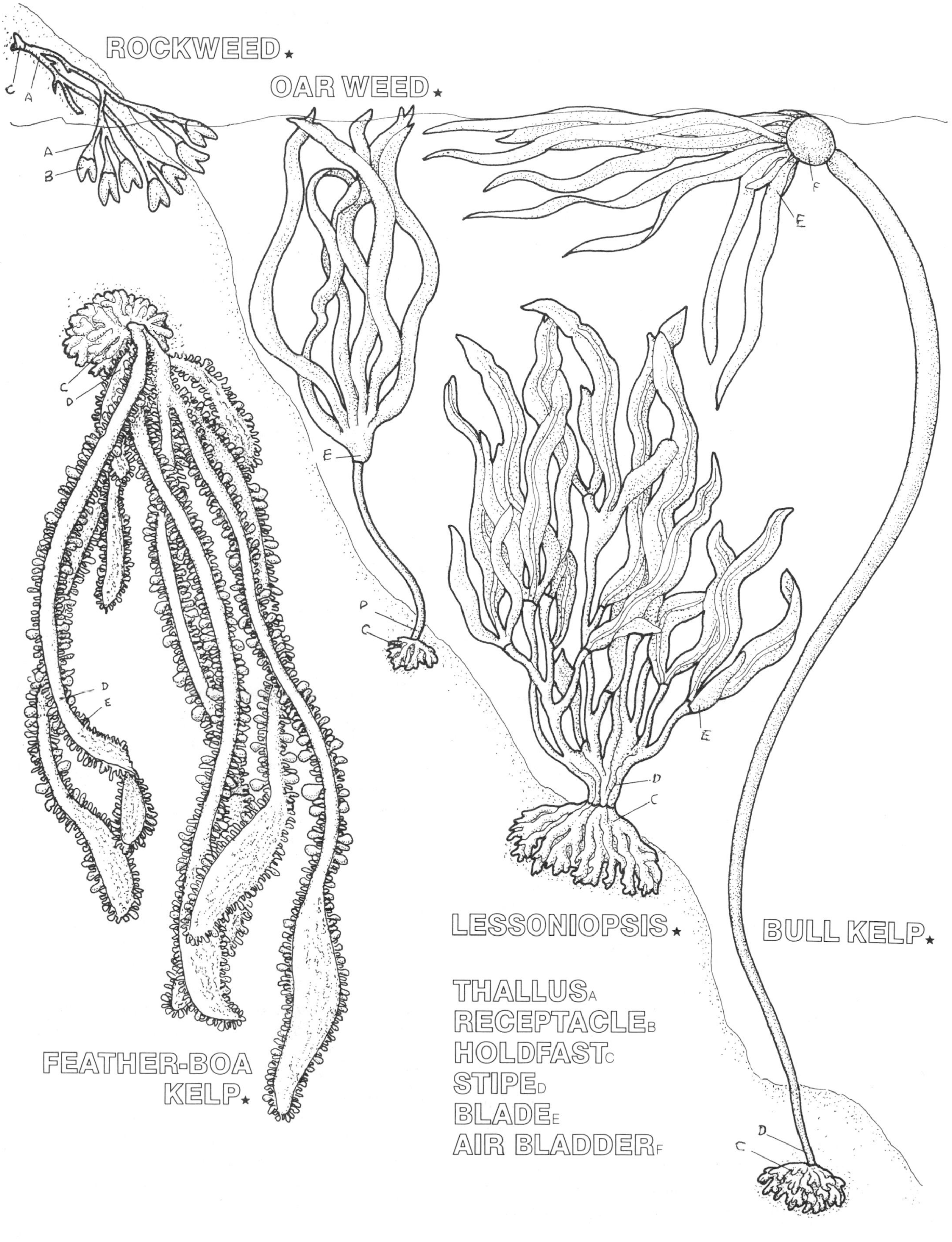

ROCKWEED ★

C A

A
B

OAR WEED ★

E

D
C

F
E

E

D
C

LESSONIOPSIS ★

BULL KELP ★

C
D

FEATHER-BOA KELP ★

C
D

D
E

THALLUS A
RECEPTACLE B
HOLDFAST C
STIPE D
BLADE E
AIR BLADDER F

Sponges (phylum Porifera: "pore bearer") are considered to be among the simplest living organisms. Almost all of the 5,000 species are marine, and all operate as filter feeders, collecting small particles of food from the sea water flowing through their bodies.

Begin by coloring the cutaway diagram in the upper right-hand corner, which shows the circulation of water through a generalized sponge. Color the body wall and the ostia. The size of the pores has been exaggerated in the drawing; they are actually microscopic in a living sponge. Next, color the filter chambers, the atrium, and the osculum.

The *body wall* of the sponge is perforated by many small pores, or ostia (sing. *ostium*) through which water enters the sponge. The water flows through a series of *filter chambers,* then passes into the *atrium,* and finally out a larger opening called the *osculum.* Sponges rely on this flow of water for feeding, gas exchange, excretion, and often, reproduction.

Note the enlargement of the single filter chamber in the circle and color all the parts labeled. Then color the enlargement of the collar cell. Finally, color the enlargement of spicules.

The heart of the sponge's water system is the *filter chambers,* which are lined with *collar cells* (choanocytes). Each individual *collar cell* possesses a single whiplike *flagellum* that beats in a rhythmic fashion. The independent rhythmic beating of many flagella creates a positive pressure within the *atrium* of the sponge, which forces water out the large *osculum* and pulls water in through the *ostia.* As the *water flow* passes over the *collar,* small particulate matter becomes trapped where it can be engulfed by the *cell body* and then digested in the food vacuoles. Larger particles of food may be engulfed by mobile cells at the entrance to the *filter chamber.*

Sponges vary in size from tiny lumpish forms to massive vaselike structures. Small skeletal elements called *spicules* are embedded in the *body wall* of the sponge and support its structure. In most sponges, the *spicules* are scattered individually in the *body wall,* as in this illustration. In the glass sponge, however, and some others, the *spicules* are organized into an elaborate, latticework skeleton. Instead of, or in addition to, the *spicules* (siliceous or calcareous), some sponges have fibers (not shown) of a protein called spongin.

The morphology of sponges varies from very simple tubular forms, to the *filter chamber* system (as shown), to complex systems involving more infolding of the *body wall* and the proliferation of smaller and more numerous *filter chambers.* The increased number of *filter chambers* allows more water to be filtered through the sponge; a 10 cubic centimeter sponge is capable of filtering 20 liters of water in 24 hours.

Color the four different types of sponges. The body of each sponge can be colored either the same color as the diagram above them, or the natural color as given in the text.

The form of sponges is influenced greatly by the available space, type of substratum, and the strength of the water movement. Most sponges are attached to hard substrata in relatively shallow water.

The purple encrusting sponge is found low in the rocky intertidal, often in large patches. This sponge may grow to 2.5 centimeters thick, and its *oscula* are quite large. The encrusting sponge does not grow tall in areas of heavy wave action that would quickly tear and destroy it; in quiet waters the *oscula* are raised on elevated craterlike projections of *body wall.*

Not all encrusting sponges attach to inanimate substrata. The smooth pink pecten sponge is found on the shells of scallops — a mutually beneficial relationship. In exchange for the substratum, the sponge covers the scallop with its porous, yielding body, offering some protection from sea star predators.

In the quiet water of subtidal habitats, such as the coral reef, large sponge forms flourish. The azure blue tubular sponge grows very tall.

The boring sponge burrows into the shells of abalones, oysters, and other molluscs. This yellow sponge lives in the tunnels it chemically etches out of the shell. Its tunneling can be extensive, severely weakening the shell. Some species of boring sponge attack corals and are responsible for much decomposition of coral reefs.

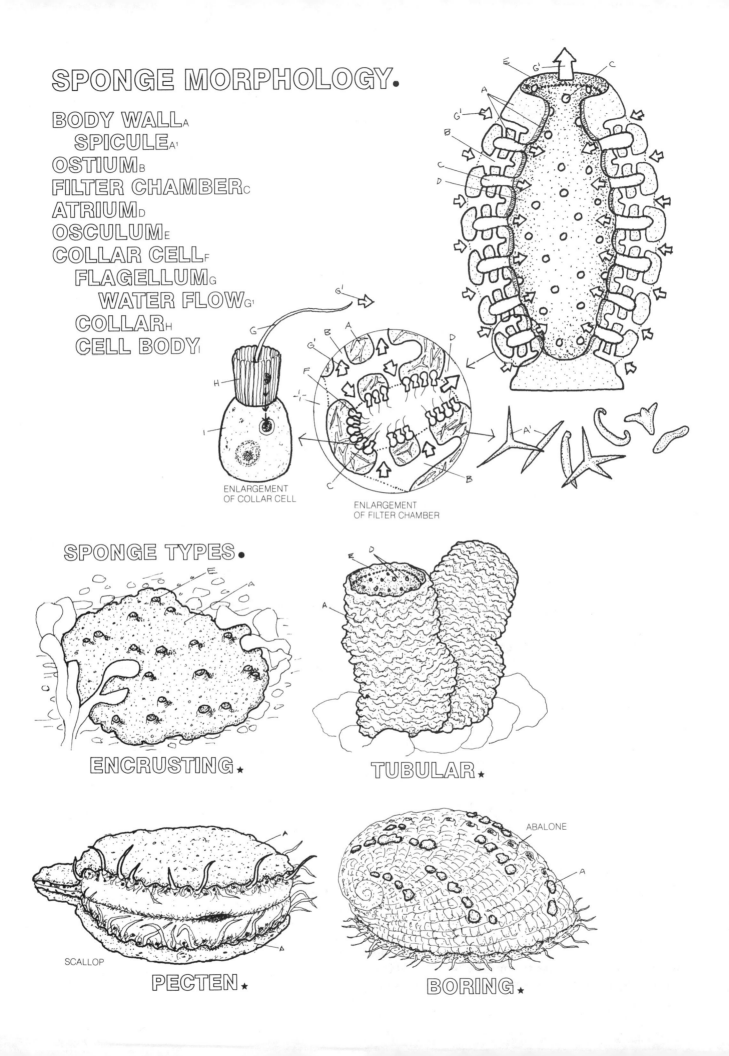

SPONGE MORPHOLOGY.

BODY WALL A
SPICULE A1
OSTIUM B
FILTER CHAMBER C
ATRIUM D
OSCULUM E
COLLAR CELL F
FLAGELLUM G
WATER FLOW G1
COLLAR H
CELL BODY I

ENLARGEMENT
OF COLLAR CELL

ENLARGEMENT
OF FILTER CHAMBER

SPONGE TYPES.

ENCRUSTING ★

TUBULAR ★

SCALLOP

PECTEN ★

ABALONE

BORING ★

16
COELENTERATE DIVERSITY: POLYPS

The phylum Coelenterata ("hollow gut") is a group of animals with a very simple, functional body structure. The digestive tract of the coelenterate lacks a second, or anal, opening, and consists of only a mouth and a saclike cavity (the coelenteron).

Begin your coloring with the diagrammatic drawings of the polyp and medusa, using a light color for the body.

There are two basic coelenterate body forms: the free-swimming medusa and the stationary (sessile) polyp. Both types have a radially symmetrical organization with the *coelenteron* located in the center. The basic difference between polyp and medusa forms is this: the medusa floats free of the substratum, with its *mouth* and *tentacles* facing downward; the polyp is attached by its *pedal disc* to a substratum, with its *mouth* and *tentacles* facing upward.

Now color the enlargement of the nematocysts.

The coelenterate *mouth* is surrounded by a ring of *tentacles* where highly specialized cells are located. These cells contain *nematocysts*—small stinging, whiplike structures that are discharged from the cells in response either to outside chemical or mechanical stimuli or direct nerve stimulus.

When potential prey makes contact with the *tentacles* of the polyp, the *nematocyst*-bearing cells are stimulated, causing the *nematocyst* to rapidly uncoil and, in some cases, penetrate the victim. Many *nematocysts* contain a venomous liquid that subdues the prey; some types of *nematocysts* are barbed or sticky, and some types actually wrap around the prey. When the prey is subdued, the *tentacles* maneuver it into the *mouth* of the polyp, and it is digested in the *coelenteron*. Undigested parts are regurgitated back out of the *mouth*.

Both polyp and medusa are known as passive predators: the polyp waits for prey to wander into its deadly tentacles, and the medusa trails its tentacles in the water as it floats along, catching its food.

Now color the different polyp types, beginning with the sea anemones, then the coral polyp, and finally, the hydroid. Note that the mouths of the colonial hydroids do not show in the drawing.

The familiar sea anemone of rocky shores is a large, single polyp. Its body is basically cylindrical in shape, with the *oral disc* at the top end, and the *pedal disc* anchoring the anemone to a solid substratum at the other end. Some species use special muscles, combined with mucous secretions, to insure a tight seal against the substratum. Sea anemones are generally attached to a solid surface, although some species prefer a burrowing existence in sand or mud, and others attach themselves to the shells of other animals. Anemones vary in size from a few centimeters in diameter to animals whose *oral discs* are 30 centimeters or more in diameter.

The squat anemone in the illustration is the giant green anemone of Pacific coast tidepools. This anemone grows to 25 centimeters or more in diameter and is capable of compressing its *body* to just a few centimeters in height. It feeds on nearly any organisms that are washed or swim unaware into its *tentacles*.

The long-columned sea anemone in the illustration may reach a height of 30 centimeters, and is usually found subtidally. Its numerous fine *tentacles* reach into the current to capture small organisms carried by the moving water.

The coral polyp is similar in structure to the sea anemone but is usually much smaller (0.5 mm–10 mm in diameter). Each coral polyp possesses a calcium carbonate *skeleton* into which the entire polyp can retract. The cuplike *skeletons* are secreted through the polyp's epidermis, and are the basic structural units that form the tropical coral reefs. Like hydroid polyps, reef corals are found in colonies; these can grow to massive sizes in an astounding variety of shapes and forms.

The polyps of marine hydroids occur generally in colonies, forming branched structures that may be attached to various substrata. An individual polyp is usually quite small (0.5–10 mm) and is specialized for a particular function: feeding, reproduction, or defense. Those illustrated here are all feeding polyps, termed hydranths, having a ring of *distal tentacles* around the mouth (not shown) and a second ring of *tentacles* around the *body,* in which the *coelenteron* is located.

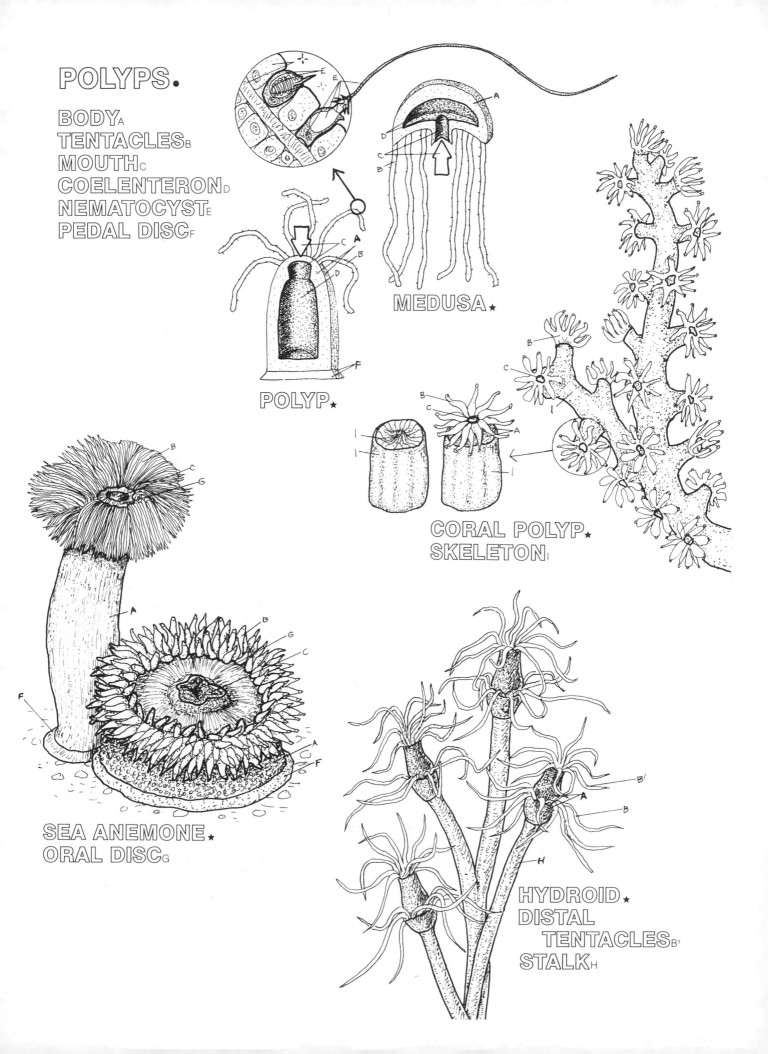

POLYPS.

BODY A
TENTACLES B
MOUTH C
COELENTERON D
NEMATOCYST E
PEDAL DISC F

MEDUSA ★

POLYP ★

CORAL POLYP ★
SKELETON I

SEA ANEMONE ★
ORAL DISC G

HYDROID ★
DISTAL
TENTACLES B'
STALK H

COELENTERATE DIVERSITY: MEDUSAE

The medusa represents the unattached, motile stage in coelenterates and is found swimming freely in the water column. Swimming is accomplished by muscular contraction of the dome-shaped *umbrella,* or bell, forcing water out and propelling the medusa in the opposite direction. Medusae remain largely at the mercy of the prevailing currents and therefore belong to the group of drifting pelagic organisms known as plankton (see Plate 10).

In coloring this plate, use the same colors as you did for similarly named structures on the previous plate. Note that the umbrella receives the same light color as the body in the polyp plate. Begin with *Polyorchis* at the top of the page, and color each medusa separately as it is mentioned in the text. The tentacles have been removed from the front edge of *Polyorchis* in order to show the underside of the umbrella. In order to get the transparent effect of the umbrellar dome in *Polyorchis* and *Haliclystus,* color the inner structures first, and then apply color (A) over everything drawn and colored under the umbrella. Note that the umbrella of *Pelagia* is thick and not as transparent as those of the other medusae, so the underlying structures are not shown.

The medusa *Polyorchis* has a high-domed *umbrella* that is transparent and clearly reveals the organs within the umbrellar space. The elongate, *mouth*-bearing *manubrium* protrudes through the opening of the umbrella. It opens into the *stomach,* which itself opens into *radial canals.* Suspended from beneath these *canals* are elongated *gonads,* or reproductive organs. The long, grasping (prehensile) *tentacles* capture zooplankton and other small marine animals. *Polyorchis* is a relatively common medusa of the bays and estuaries of the west coast of the United States; specimens with *umbrellas* over 5 centimeters in height are not unusual.

The squat medusa *Aurelia* is a jellyfish with no *manubrium,* relatively short *tentacles,* and four elongated *oral arms* that surround the *mouth* opening. *Aurelia* is found, often abundantly, along both the Atlantic and Pacific coasts. Instead of entrapping prey with its *tentacles, Aurelia* is a suspension feeder. As it sinks through the water, *Aurelia* catches plankton in the mucus on the inside of its *umbrella.* The mucus-covered food is then carried to the margin of the *umbrella* and scraped off by the *oral arms.* The *arms* have ciliated grooves that carry food to the *mouth.* The *umbrellar* diameter of *Aurelia* reaches 15 centimeters.

Another typical jellyfish is *Pelagia.* This animal often exceeds 75 centimeters in *umbrellar* diameter and occurs in large numbers along the west coast of the United States. The four *oral arms* are very maneuverable. A large *Pelagia* trails its nematocyst-bearing *oral arms* as much as 2.5 meters below the *umbrella,* entrapping and subduing small organisms that swim into them. The *oral arms* contract to deliver prey to the mouth (not shown), which is located at the center of the *umbrella.* In southern California, many *Pelagia* are swept shoreward during the summer months and are the cause for a large number of the jellyfish stings experienced by bathers.

The small (2.5 cm) jellyfish *Haliclystus* does not drift or swim freely, as do most medusae. Instead, *Haliclystus* attaches itself to surf grass in the rocky intertidal zone, and to eelgrass in quiet waters. Its small, stalked *attachment disc* protrudes from the center of the upturned *umbrella.* The *gonads* radiate out from the center and give the medusa the appearance of a webbed basket adorned with eight *tentacle* clusters. The mouth (not shown) is at the center of the *umbrella. Haliclystus* feeds on small planktonic organisms and other small animals living on the plant to which the medusa is attached. Some species of *Haliclystus* are able to absorb pigment from the plant to which they attach, thus changing the color of their *umbrella* to match their substratum.

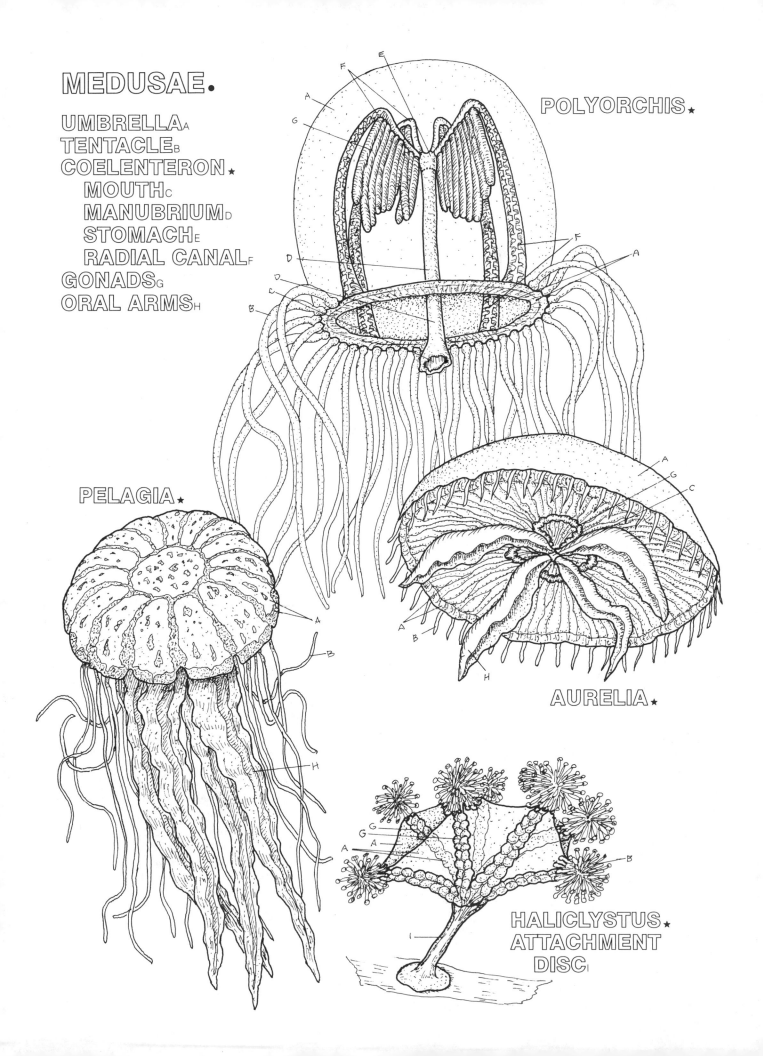

MEDUSAE.

UMBRELLA A
TENTACLE B
COELENTERON ★
MOUTH C
MANUBRIUM D
STOMACH E
RADIAL CANAL F
GONADS G
ORAL ARMS H

POLYORCHIS ★

PELAGIA ★

AURELIA ★

HALICLYSTUS ★
ATTACHMENT
DISC I

18
MARINE WORM DIVERSITY: COMMON WORMS

The marine worms include several groups of animals that occur in a great variety of sizes, shapes, and colors; many are quite agile and graceful in the water.

Color each worm as it is discussed in the text. Begin with the flatworm, and notice that the pharynx, located on the under or ventral side is visible from the top through the thin body.

The free-living flatworms range in size from almost microscopic to 60 centimeters in length. The great majority of the 3,000 known species are marine, and most of these are bottom dwellers, living in sand or mud under rocks and algae. The marine flatworm illustrated here depicts a generalized form (such as *Notoplana*), and is commonly found in the rocky intertidal zone of the west coast of the United States. It lives on the underside of large boulders, and can be seen gliding along the rock, by means of the cilia on its ventral surface. *Notoplana* has two dark *eyespots* at the anterior of its brownish-gray *body*. The dark area located on the midline of the *body* is the only opening of the digestive tract. What appears to be a ruffled curtain is the retracted *pharynx,* which can be extruded for food gathering (shown in the illustration). The flatworm shown here is a nocturnal predator, feeding on small molluscs, crustaceans, and other invertebrates.

The ribbon worms (over 600 species) are closely related to flatworms. They are capable of tremendous lengthwise extension; a worm that measures 20 centimeters when contracted can stretch to over a meter! The ribbon worm shown here lives in a parchment-like tube among the algae, mussels, and other organisms on low intertidal and subtidal rocks and pilings along the west coast.

Unlike the flatworms, ribbon worms possess a complete digestive tract (both mouth and anus) and also have a food gathering device called an eversible *proboscis,* which is generally separate from the digestive system. When not in use, the *proboscis* is retracted inside a cavity above the digestive tract. The *proboscis* can be everted (shot out) anteriorly to coil around the worm's prey. A very sticky mucus is secreted from the *proboscis* to aid in the capture of the prey. In some ribbon worm species, the *proboscis* may be as long or longer than the worm itself, and may be equipped with a piercing barb or stylet and poison glands. Ribbon worms are generally nocturnal carnivores, feeding on other worms, molluscs, crustaceans, and small fish.

The peanut worms, or sipunculids, are a group of about 300 species. They are deposit feeders living burrowed in mud or sand flats, in muddy crevices between rocks, in coral crevices, in abandoned shells of gastropods, or in the tubes of polychaete worms. Peanut worms range in size from 0.2 to 72 centimeters; the average length is about 10 centimeters. Their bodies consist of two basic sections: the rounded, bulbous *trunk* and the narrower *introvert*. The *introvert* is the anterior portion of the worm's body, and can be retracted into the *trunk*. The mouth is at the tip of the *introvert* and is surrounded by ciliated *tentacles* that are used in feeding. A peanut worm with a 5-centimeter *trunk* can extend its *introvert* out 15 centimeters in search of food, while it sits safely in a crevice.

The incredible Galapagos tube worm, recently discovered on the floor of the Pacific Ocean, often measures over 3 meters in length. Large colonies of these worms have been found in deep ocean areas where warm water seeps out of the earth. The Galapagos tube worm has no mouth, and it was first speculated that the worms extracted dissolved materials from the water with their red *tentacles*. However, more recent studies suggest that the tube worm may harbor bacteria in its tissues that derive energy from the chemical reduction of mineral substrates. The tube worms may be deriving their nourishment from these chemo-autotrophic bacteria.

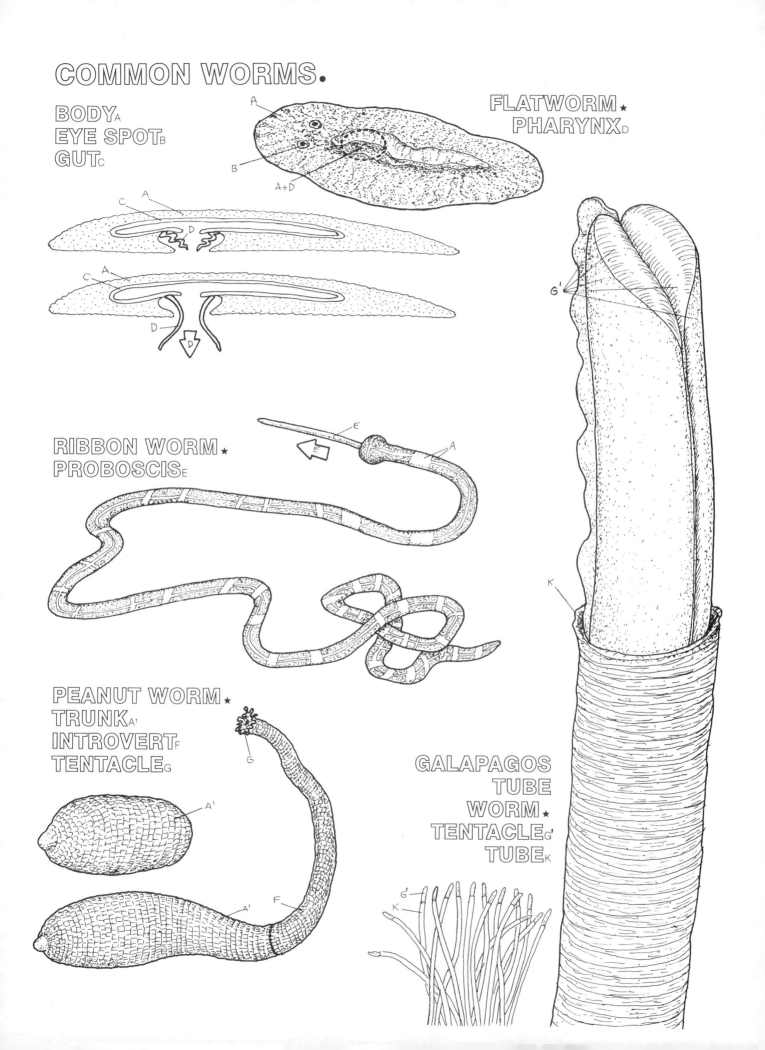

COMMON WORMS.

BODY_A
EYE SPOT_B
GUT_C

FLATWORM ★
PHARYNX_D

RIBBON WORM ★
PROBOSCIS_E

PEANUT WORM ★
TRUNK_{A¹}
INTROVERT_F
TENTACLE_G

GALAPAGOS
TUBE
WORM ★
TENTACLE_{G¹}
TUBE_K

MARINE WORM DIVERSITY: INNKEEPER WORM

The fat innkeeper worm is an interesting marine worm that inhabits semi-permanent burrows in sandy mud flats along the California coast. Normally about 20 centimeters long, this rotund worm possesses a highly stretchable (distensible) *proboscis* and a pair of gold-colored, hooked *bristles* near its anterior end. A circle of similar *bristles* surrounds its *anus*.

Color the drawing of the innkeeper worm a medium pink, which is its true color. Then color the diagram of peristaltic action. Here, the worm receives two colors, one for its body and one for the area of muscular contraction (E). Be sure to color the arrows that indicate the direction of the peristaltic contraction (E) and water flow (F), using a light blue for the water.

The innkeeper worm makes its *burrow* by digging with its *proboscis* and its anterior hooked *bristles*. The *anal bristles* and *body* movement carry the material backward and out of the *burrow*. Once the *burrow* is constructed, the innkeeper stays there unless disturbed.

Now, color the worm in its mud flat habitat. Note that the sea water above the burrow and the sea water in the burrow receive the same color, indicating that the burrow is flooded. Color the mud light brown. Color the proboscis of the worm in the burrow, then color over it with the light color chosen for the mucous net.

The innkeeper worm pumps water through its *burrow* by means of *peristaltic action*. Waves of *muscular contraction* create a "bottleneck" whose travel down the worm's *body* pushes water before it and eventually out the *burrow*. To feed, the innkeeper secretes a *mucous net* placed snug against the *burrow*

walls near one opening. The worm backs down the *burrow*, secreting the *net* as it goes. When the *net* is between 5 and 20 centimeters long, the innkeeper stops, positions itself in the *burrow*, and begins drawing water through the very fine mesh of the *net*, using the peristaltic action just described. When the *net* is laden with food particles, the worm moves back up the *burrow*, eating both the *net* and small particles of food; large chunks are discarded. Thus, the innkeeper is a filter feeder whose filter is outside its own *body*.

Color the other marine animals in the innkeeper's burrow, noting their positions in the burrow. Also color the larger illustrations of these animals.

The large food particles discarded by the innkeeper worm do not go to waste. There are several marine animals that share the innkeeper's home and take advantage of both the extra availability of food and the security of the *burrow*.

The small (2 cm) *pea crab* and the polychaete *worm* (4 cm) share the *burrow* and fight over the food discards. The polychaete *worm* remains in contact with the innkeeper in order to gain an advantage over the quicker *pea crab*.

Two other frequent residents of the *burrow* are a *goby* and a small *clam*. The *goby* uses the *burrow* as a home base and forages for its own food out on the mud flat at high tide. The *clam* (1.7 cm) is the innkeeper's fourth guest. Instead of making its own shallow *burrow* at the surface of the mud flat where it might be washed away or eaten, it digs into the wall of the innkeeper's *burrow* some distance below the opening. From here, the *clam* extends its short siphons into the *water currents* flowing through the large *burrow*, and from these currents it is able to siphon food.

INNKEEPER WORM.

PROBOSCIS_A
BODY_B
ANUS_C
BRISTLES_D

PERISTALTIC ACTION ★
MUSCULAR CONTRACTION_E
WATER CURRENT_F

MUD FLAT HABITAT ★
MUCOUS NET_G
SEA WATER_{F1}
   BURROW_{F2}
MUD_H
BURROW GUESTS ★
   GOBY_I
   CLAM_J
   PEA CRAB_K
   WORM_L

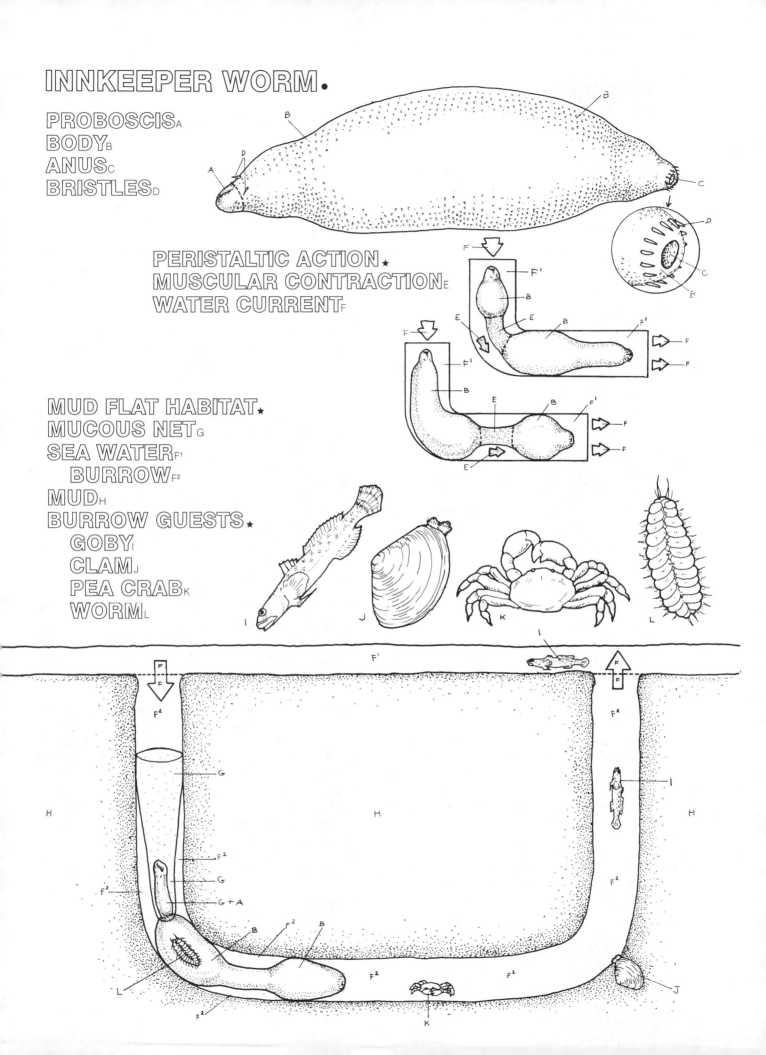

Of the many phyla of worms, the segmented or annelid worms are the most diverse and perhaps the most beautiful. This group includes the familiar earthworm, the leech, and, in the marine environment, the class of polychaetes, comprising over 5,000 species.

The annelid body is divided by partitions into compartments (segments) that, in part, restrict the flow of body fluids. This segmentation enables the burrowing annelid to dig much more efficiently than the non-segmented worm. In this plate a non-burrowing polychaete (Nereis) and two burrowing polychaete worms will be introduced.

Color Nereis and the enlarged views of its head region at right. Nereis is often an iridescent blue-green. The parapodia and their setae should be colored in the enlarged view of Nereis; the setae need not be colored on the drawing of the complete worm.

The clam worm, Nereis, is a widely distributed genus and is minimally specialized. Nereis has what may be considered the typical polychaete body, consisting of repeated identical body segments, each with a pair of lateral paddle-like appendages called parapodia. The parapodia are flattened projections of the body and are equipped with rods called setae. The setae project through the parapodia and are connected to muscles that enable them to retract or extend. As Nereis crawls along in serpentine fashion, the setae aid in gripping the substratum.

The head of Nereis consists of two segments, a prostomium and a peristomium. The prostomium is positioned in front of the mouth and bears several sensory structures; these include the light-sensitive eyes, as well as the antennae and palps, which appear to be receptors for both chemical and tactile senses. The peristomium, just behind the prostomium, contains the mouth and four pairs of tentacular cirri that also act as tactile receptors. The concentration of sensory structures in the head area means that Nereis receives information about its environment as it moves into it.

The mouth of Nereis harbors an eversible proboscis. The proboscis remains folded in on itself until contracting body wall musculature increases pressure on the body fluid, which, in turn, everts the proboscis. The proboscis is armed with jaws that swing open and then clamp shut as the body fluid pressure is reduced and the retractor muscles pull the proboscis back in. Various species of Nereis are found moving about freely in many habitats, including the rocky intertidal zone and mud and sand flats.

Now color Glycera and note its everted proboscis. The natural color of Glycera is a dark pink or light red. Also color the smaller drawing that shows the worm in its network of tunnels. Give the flooded galleries and the water above them a light blue color. Color the sand a light gray.

Glycera is a sand-flat-dwelling carnivore possessing a proboscis that is one-fifth its own body length and armed with four stout horn-like jaws, each with its own poison gland. Glycera constructs a nest of interconnecting burrows (galleries) in the substratum with many openings to the surface of the sand flat.

Glycera's prostomium is conical and adorned with four short antennae. It is sensitive to changes in water pressure, such as caused by the movement of prey above the nest. Glycera feeds on polychaetes and other invertebrates. Some Glycera reach a length of more than 50 centimeters.

Next, color Arenicola (the "lugworm") and its burrow environment. The arrows indicating the movement of sand and water should also be colored. The natural color of Arenicola ranges from pink to dark green.

Arenicola also lives on sand flats, but it is a deposit feeder, not a carnivore. It passes sand through its digestive tract by means of peristaltic action and removes the organic matter from the sand. Arenicola's burrow is easily identified by the pile of fecal mounds located at one end. The worm's feeding and burrowing activities aerate the upper layers of the sand flat, providing circulation and re-exposure of buried sediments and the nutrients they contain.

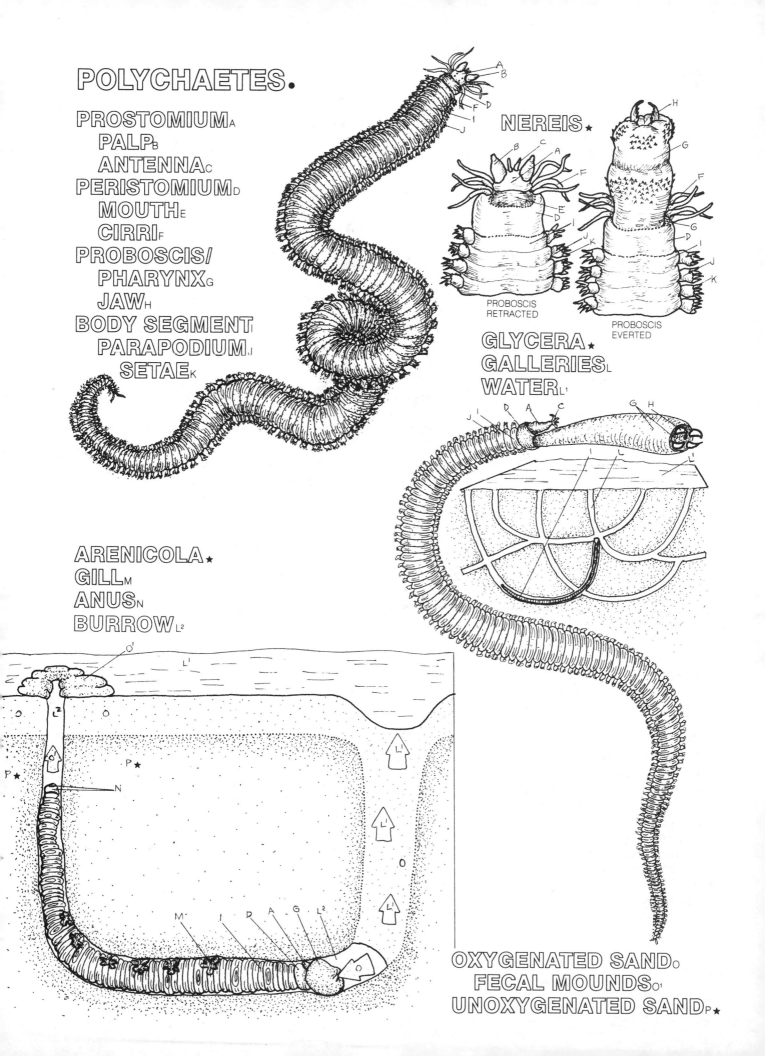

POLYCHAETES.

PROSTOMIUM_A
PALP_B
ANTENNA_C
PERISTOMIUM_D
MOUTH_E
CIRRI_F
PROBOSCIS/
PHARYNX_G
JAW_H
BODY SEGMENT_I
PARAPODIUM_J
SETAE_K

NEREIS ★

GLYCERA ★
GALLERIES_L
WATER_L1

PROBOSCIS
RETRACTED

PROBOSCIS
EVERTED

ARENICOLA ★
GILL_M
ANUS_N
BURROW_L2

OXYGENATED SAND_O
FECAL MOUNDS_O1
UNOXYGENATED SAND_P ★

21

MARINE WORM DIVERSITY: TENTACLE-FEEDING POLYCHAETES

Free-living polychaetes, such as *Nereis,* actively seek out their food and shelter. Other types of polychaetes remain stationary, living in tubes, burrows, or crevices and obtaining food by extending their *tentacles* out into the water. Two ways of obtaining food with tentacles are exemplified by the fan worm, *Sabella,* and burrowing polychaete, *Amphitrite.*

Color the fan worm in the upper left. You may wish to use a bright color for the mass of radioles, as they are very colorful in life. Next, color the enlargement of the tube and radioles. The tube is opaque, but for purposes of illustration it is shown as transparent so you can color the body segments of the worm as it resides in its tube.

Fan worms are conspicuous members of wharf piling and coral reef communities. Their beautiful flowerlike fan is actually a group of tentacle-like projections called *radioles,* extending from the prostomium. The *radioles* have lateral branches, or *pinnules,* covered with fine cilia (not shown). When the fan worm emerges from its parchmentlike *tube,* the *radioles* spread in a funnel-shaped crown around the *mouth.* The cilia beat in unison and create a current, bringing water up through the crown of *radioles* and out the top. *Food particles* carried in the current are trapped and moved down the *pinnule* to a *food groove* that runs the length of the *radiole.*

Now color the diagrammatic enlargement of the radiole and pinnules.

The ciliated *food groove* carries food toward the *mouth*; at the base of the *radiole* the various *food particles* are sorted by size. Large *particles* are rejected; special ciliary tracks on the *palps* carry these back into the excurrent stream, flowing up and out the center of the crown of *radioles.* The smallest *particles* are carried to the *mouth* and consumed. The medium *particles* are carried to a special *ventral sac,* where they are stored for use in tube-building.

At the base of the *radiole* crown is a fleshy fold of tissue called the *collar,* which holds the fan worm securely at the top of its *tube.* Mucus secreted from both the *collar* glands and the glandular *ventral plate* is mixed with the medium-sized *particles* stored in the *ventral sac* to create a thin thread of *tube* material. The fan worm rotates slowly, and as it lays down this material along the edge of its papery but flexible *tube,* gradually repairs and lengthens it.

Color the tentacles of *Amphitrite,* as well as its gills and all of the body segments. In the drawing at lower right, note that it illustrates the three ways a tentacle can transport a food particle to the region of the mouth, where the particle is sorted and either eaten or rejected.

The polychaete *Amphitrite* lives in semi-permanent burrows on mud flats or in rock crevices in the intertidal zone; it feeds on organic material deposited on the surface of the substratum. *Amphitrite* possesses a multitude of hollow, distensible, ciliated, prostomial *tentacles* that reach out over the substratum until a *food particle* is found. *Tentacles* move *food particles* to the *mouth* one of three ways, depending on size. If the *particle* is very small, the *tentacle* forms a shallow ciliated food groove that carries the food to the *mouth.* For somewhat larger *particles,* the action of the ciliated food groove is augmented by peristaltic contraction along the length of the *tentacle.* With very large *particles,* the *tentacle* wraps around it and is retracted into the *mouth,* carrying the *particle.* Each *tentacle* is pulled separately through the folded upper *lip,* which sorts by size and rejects the *particle* or pushes it into the *mouth.*

Bright red *gills* filled with blood are located near the head region of the worm. Both the *gills* and the *tentacles* may be injured or removed by predators, but the worm is capable of regenerating these structures. The elastic *tentacles* of *Amphitrite* can reach out over a wide area: a worm 3 to 4 centimeters in length can cover a diameter of 18 to 20 centimeters with a radiating maze of *tentacles.*

TENTACLE-FEEDING POLYCHAETES.

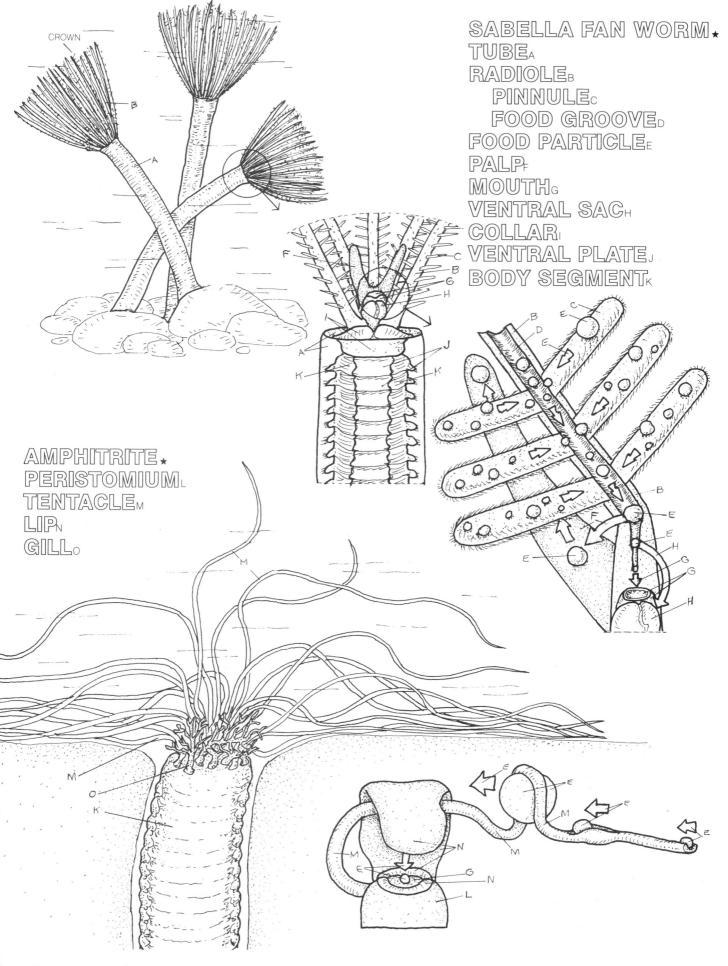

CROWN

SABELLA FAN WORM.★
TUBE A
RADIOLE B
PINNULE C
FOOD GROOVE D
FOOD PARTICLE E
PALP F
MOUTH G
VENTRAL SAC H
COLLAR I
VENTRAL PLATE J
BODY SEGMENT K

AMPHITRITE.★
PERISTOMIUM L
TENTACLE M
LIP N
GILL O

22

MOLLUSCAN DIVERSITY: INTERIOR OF A SHELL

A common introduction to marine biology is through the diversity and beauty of sea shells. Although these shells are interesting to collect and examine, the molluscs that once lived within are even more fascinating. This plate provides an understanding of how some molluscs are structured and how they function, using the clam called a cockle as an example.

Start by coloring the shell interior at the upper right. Use light colors for the shell and mantle.

On the empty right shell, or *valve,* shown here, is a peak where the shell began its formation. Called the beak, or *umbo,* of the *valve,* this peak can be used as a reference to indicate the dorsal side (back) of the clam. Below and to the right of the *umbo* (as pictured), is the *hinge ligament.* Made of protein, this compressible structure serves to connect the *valves* and functions in the opening and closing of them. Below and to the side of the *ligament* are a number of projections called *hinge teeth,* which fit into corresponding recesses (sockets) in the other *valve.* This tooth and socket arrangement aids in the articulation of the *valves* by preventing one from riding over the other. This is important in burrowing, or when the clam is being attacked by a predator, since a tightly closed shell is its most effective defense.

Also visible in the empty *valve* are four oval *muscle scars.* These *muscle scars* are the sites of attachment of the *adductor muscles,* which pull the *valves* together and hold them shut, and of the pedal (foot) retractor muscles (not shown). The thin, curving line joining the two *adductor muscle scars* is called the *pallial line,* and it marks the point where the fleshy *mantle* attaches to the *shell.*

Color the side view and the cross section of the two valves on the left of the plate. In the cross-sectional view, the cut is made through a single adductor muscle. The visceral mass has been removed. Also color the arrows that indicate the direction the valves move as they close upon contraction of the muscle.

Looking at the two *valves* in the cross-sectional view, one can see that the *adductor muscles* and the *ligament* have opposing roles. When the *adductors* contract and the *valves* are brought together, a portion of *ligament* is compressed and a portion is stretched.

When the *adductor muscles* relax, the compressed part of the *ligament* expands, and the stretched upper part of the *ligament* contracts. This results in the *valves* gaping open so the clam can extend its *foot* and *siphons.*

Also visible in this cross-sectional view is the fleshy *mantle* which completely underlies the *valve* and is responsible for its secretion and maintenance.

Color the clam at the top of the page and the internal view of the clam at the bottom of the page. Note that some names refer to related structures seen in the empty shell. Arrows that indicate the direction of the feeding current (below the gills) should be colored the same as the incurrent siphon (I). Those arrows above the gills should be colored the same as the excurrent siphon(J).

At the top of the plate, the cockle is shown in a side view with its *incurrent* and *excurrent siphons* extended from the posterior end, and its large *foot* extended from the anterior end. This is the normal position of a burrowed cockle while actively pumping water for feeding and respiration. The radial ridges on the outside of the shell add strength and help anchor the cockle in the sand.

The illustration of the cockle at the bottom of the plate shows the right *valve* with its underlying *mantle* removed to expose the internal organs. Cut free from the left *valve,* the large *adductor muscles* are clearly visible. It can also be seen how the extended *siphons* are continuous with the fleshy *mantle* that lines the inside of the *shell.* Note the large bilobed *gill,* the smaller *labial palp,* and the *foot* beneath the *gills.* These *gills* are covered with microscopic, hairlike processes called cilia (not shown), which wave in unison to create a current. The arrows indicate the direction of the water current. As water passes through the *gills,* small particles, such as phytoplankton, are filtered out and are carried toward the mouth (not shown) on mucous strands by specialized cilia. Deposited onto the *labial palps,* the particles are sorted by size. Smaller particles are directed to the mouth, located beneath the *labial palp.* Rejected, larger particles accumulate below the *gills,* near the *foot,* and are periodically expelled.

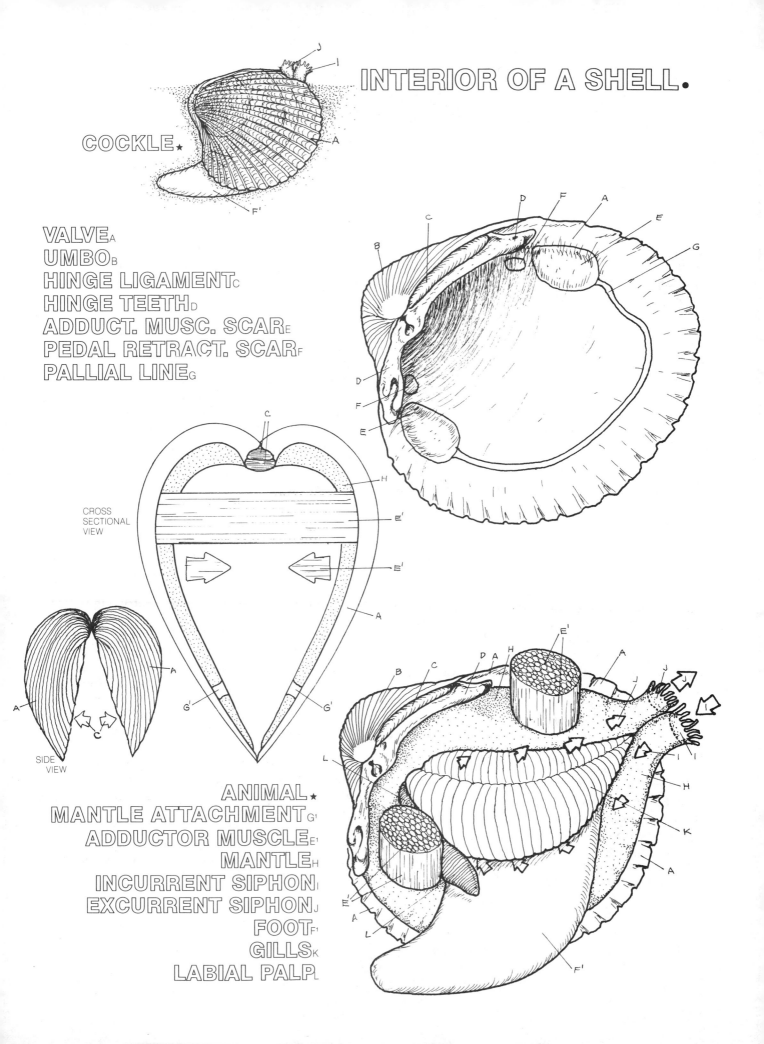

INTERIOR OF A SHELL.

COCKLE.★

VALVE ᴀ
UMBO ʙ
HINGE LIGAMENT ᴄ
HINGE TEETH ᴅ
ADDUCT. MUSC. SCAR ᴇ
PEDAL RETRACT. SCAR ꜰ
PALLIAL LINE ɢ

CROSS
SECTIONAL
VIEW

SIDE
VIEW

ANIMAL.★
MANTLE ATTACHMENT ɢ¹
ADDUCTOR MUSCLE ᴇ¹
MANTLE ʜ
INCURRENT SIPHON ɪ
EXCURRENT SIPHON ᴊ
FOOT ꜰ¹
GILLS ᴋ
LABIAL PALP ʟ

23

Variations in the structure and shape of bivalve molluscs reflect evolutionary adaptations to different environments. Most malacologists (scientists who study molluscs) concur that the bivalve group evolved from organisms originally adapted to living in soft sediments, such as sand or mud, and that some secondarily took up existence on top of the substratum. The discussion that follows includes three soft-substratum dwellers (members of the infauna) — the cockle, the softshell clam, and the bent-nosed clam; and two bivalves, the mussel and the scallop, that have adapted to living on top of the substratum (epifauna).

Color separately the parts of each bivalve as the animal is discussed in the text. Use a color for the shells light enough that the texture of the surface will show. Use a light color for the scallop's mantle and then dot the "eyes" with a contrasting darker color.

The cockle is found living very close to the surface, primarily in sandy substrata. Its *siphons* are very short; the *incurrent* and *excurrent siphons* face slightly different directions to insure that the same water is not refiltered. Because the cockle lives so close to the surface, it is often exposed or dislodged by water movement. The cockle's large digging *foot* is extremely useful in reburrowing and escaping from predators.

The softshell clam lives typically in very soft sandy mud, where it burrows very deeply. For filter feeding, it has elongated *siphons* that enable it to reach beyond the surface, into the water current. As the clam grows larger, it burrows deeper, and its *siphons* lengthen accordingly. The softshell clam is rarely dislodged from its deep burrow; therefore its need for rapid burrowing ability is much less than that of the cockle, and its *foot* is much smaller.

The bent-nosed clam is not a filter feeder, but a deposit feeder, using its long, maneuverable *incurrent siphon* to probe along the surface of the sediment for deposited organic matter. The clam is named "bent-nosed" for the noticeable curve in the posterior region of its *shell*. The clam lies in the sediment on its left side with its bent-nosed posterior curved upward, and its *incurrent siphon* extended beyond the muddy surface. Food is sucked up through the *incurrent siphon,* carried into the mantle cavity, and collected on the gills (not shown). Unlike its filter-feeding cousins who can remain relatively stationary as the water brings their food to them, the deposit-feeding bent-nosed clam soon depletes its food supply in an area and must move through the sediment to a new position. To facilitate this movement, the bent-nosed clam is very thin and possesses a broad, thin, and maneuverable digging *foot*.

The edible mussel lives on the substratum attached to pier pilings or rocks in the intertidal and shallow subtidal zones. It often occurs in large aggregations called mussel clumps. Attachment to the substratum is by means of special *byssal threads,* secreted as a liquid from a gland near the *foot*. The *threads* harden upon contact with the water, serving to hold the mussel in position. The mussel lacks large protruding siphons, and instead needs only a small ruffled area at its posterior end in order to direct the inflow and outflow of water current for its filter feeding.

The scallop has developed adaptations appropriate to its motile existence. The scallop is a filter feeder that lives on the sediment surface, completely unattached to any substratum. It needs neither siphons for feeding, nor a foot for digging. It has rows of "eyes" along the edge of the *mantle* that can detect shadows and movement and aid in avoiding predators. The *eyes* may be blue, red, gold, or other colors, depending on species. Along the *mantle* edge are mantle tentacles, which are sensitive to touch and contain cells sensitive to chemical information (chemoreceptors). These aid the scallop in perceiving its environment as it swims through the water by clapping its valves together (see Plate 86).

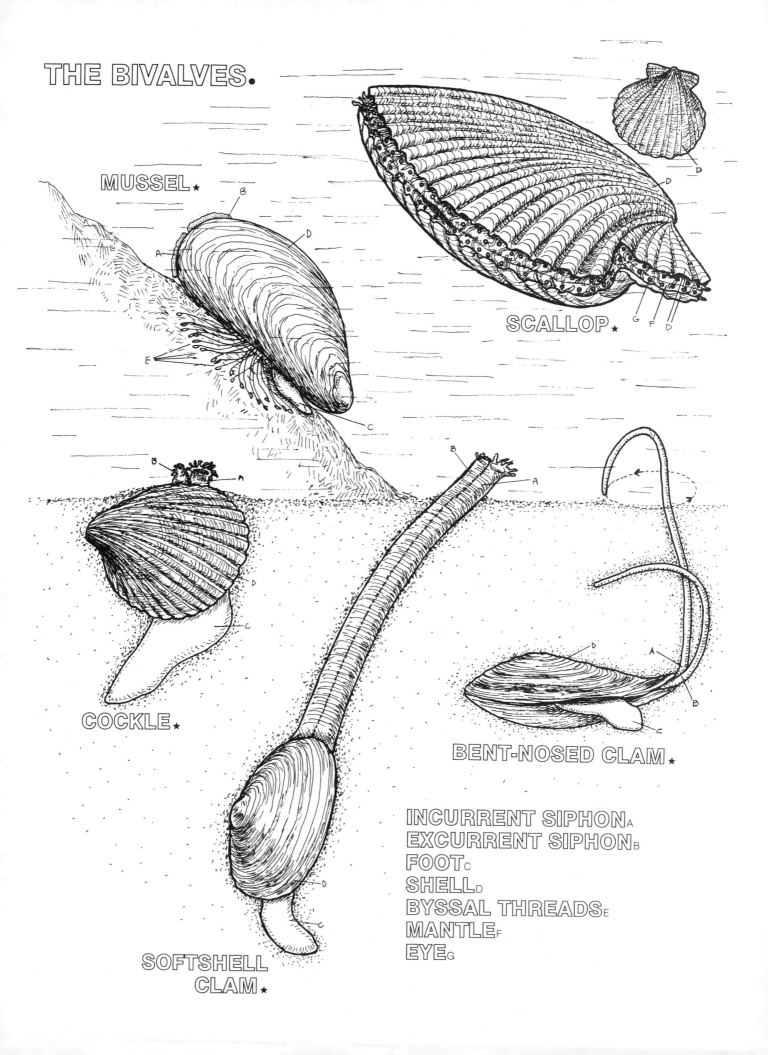

THE BIVALVES.

MUSSEL ★

SCALLOP ★

COCKLE ★

BENT-NOSED CLAM ★

SOFTSHELL
CLAM ★

INCURRENT SIPHON A
EXCURRENT SIPHON B
FOOT C
SHELL D
BYSSAL THREADS E
MANTLE F
EYE G

24

MOLLUSCAN DIVERSITY: SHELLED GASTROPODS

The gastropods (Class Gastropoda) are the largest class of molluscs: 15,000 fossil forms are identified, and existing species number well over 35,000.

Color the lower empty shell of the tulip snail and use a light color for the body whorl.

Most gastropod shells are built as a series of spirals, called *whorls.* The tip, or *apex,* of the shell, is the smallest *whorl* laid down by the snail in the early part of its life. As the snail grows, it lays down the intermediate *whorls* which form the *spire* of the shell. The final large spiral is the *body whorl,* terminating at the *aperture* or opening. The *aperture* is elongated into an anterior notch or *siphonal canal,* which harbors the incurrent respiratory *siphon* in the living snail.

Color the illustration of the living tulip snail. Use a dark color for the eyes.

The tulip snail possesses a tough, proteinaceous oval-shaped structure called the *operculum,* which is carried on its broad *foot* and is used to shut the snail snugly into its shell. When disturbed, the tulip snail retracts into its shell by first pulling in its *head,* then its *foot,* with the *operculum* brought in last to seal off the shell.

Projecting anteriorly is the elongated *siphon* used to carry water to the gills (not shown) for respiration. Special chemosensory organs are located near the gills. The *tentacles* are chemosensory and touch-sensitive; the *eyes* are light-sensitive and can detect movement.

Tulip snails are predatory molluscs, feeding on other molluscs, particularly on bivalves. The tulips reach a length of 10 centimeters and are commonly found in the Gulf of Mexico and along the southern coast of the United States.

Color the two illustrations of the abalone.

The abalone is a large herbivore common to the Pacific coast of North America. The abalone lives in shallow rocky areas of considerable wave action, and the flat shell shape offers little resistance to water movement. The broad, flat *body whorl* terminates in an aperture that is as large as the *whorl* itself. The *foot* of the abalone completely fills the aperture. Because of the size and tremendous surface area of the *foot,* the abalone can grip the substratum with amazing tenacity and remain securely fastened against strong waves and most predators.

The abalone has a pair of sensory *tentacles* and a large *mouth.* Around the *foot* of the abalone is a *mantle* from which the *epipodial tentacles* protrude. If the *epipodial tentacles* are touched, the *mantle* retracts and a strong muscular contraction of the *foot* brings the shell down tightly against the rock surface.

The abalone shell, measuring up to 37.5 centimeters in some species, has several openings through which the excurrent respiratory water flows. These *excurrent openings* also carry out the waste products of digestion and excretion and serve as the exit for sex cells when the abalone spawns (see Plate 64).

Color the cowry and the moon snail as they are discussed in the text.

Some of the most beautifully patterned shells are those of the cowries, found in tropical and subtropical oceans. The cowry shell has a glossy, polished appearance, which is maintained by the cowry's ability to completely encase its shell with its *mantle.* The two large *mantle* lobes can be drawn up the sides of the shell to meet at the dorsal midline, or can be completely withdrawn into the shell. The *mantle,* shown in the illustration only partially covering the shell, is often brightly colored and patterned and sometimes studded with small fleshy projections called papillae (not shown).

The cowry moves on its *foot,* probing with its *tentacles* and *eyes,* taking in the respiratory current through its short funnel-like *siphon.* The cowry feeds on small bottom-dwelling invertebrates such as compound sea squirts and dead animals.

The cowry shell grows up and over itself, so that in the adult animal shell only the *body whorl* is visible. Cowry species range in size from 6 to 150 millimeters.

The moon snail lives on the mud and sand flats, where a very large *foot* aids its movement through the soft substrata. Locomotion is further aided by the *propodium,* an extension of the *foot* that serves as a plow to dig forward through the mud. The *propodium* has a flap that extends to cover the head of the snail, providing protection and leaving only the *siphon* and the *tentacles* exposed as the snail travels through the sand and mud. Note the *spire* and *body whorl* in the moon snail shell. The snail preys on bivalve molluscs and other gastropods.

SHELLED GASTROPODS.

SHELL CHARACTERISTICS ★
SPIRE A
   APEX B
BODY WHORL C
APERTURE D
   SIPHONAL CANAL D'

SOFT PARTS ★
OPERCULUM E
FOOT F
SIPHON G
TENTACLE H
EYE I
MANTLE J
HEAD K
MOUTH L
EPIPODIAL TENTACLE M
EXCURRENT OPENING N
PROPODIUM O

TULIP SNAIL ★

VENTRAL
VIEWS

ABALONE ★

MOON SNAIL ★

COWRY ★

25

MOLLUSCAN DIVERSITY: SHELL-LESS GASTROPODS

The shell-less gastropods (sea slugs) are a group of marine organisms that have shed the passive defenses of the shell and operculum and instead have developed complex chemical and biological defenses to ward off their predators. They have adapted vividly colored warning patterns and are sometimes referred to as "butterflies of the sea." Illustrated and discussed here are two nudibranch sea slugs and the closely related sea hare.

Begin by coloring the dorid nudibranch as described below. If you wish, color the encrusting sponge, on which the dorid is feeding, purple or red.

Dorid nudibranchs dwell in rocky intertidal zones throughout the world, and are common along the Pacific coast of North America. The spotted *mantle* of the dorid covers the entire dorsal surface and hangs down over the *foot*. The *mantle* surface of many dorids is brilliantly colored or patterned and is thought to act in warning predators of the dorid's unpleasant taste. In a few cases the color matches the background on which the animal feeds. This dorid nudibranch is grey or light brown with dark brown or black spots. The average size is 7.5 centimeters and the large, broad *foot* and flattened body are adapted for feeding on its sole prey — the encrusting sponge.

A circlet of *gill plumes* protrudes from the dorsal surface and is capable of complete retraction into a special pocket. A pair of dorsal chemosensory tentacles, the *rhinophores,* can also be retracted into the special pocket structures.

Now, color the aeolid nudibranch as described below.

The dorsal ornamentation of the dorid pales in comparison with that of the aeolid nudibranch. The aeolid has clusters of elongated dorsal structures called *cerata,* often brightly colored with vivid contrasts, and these are thought to draw attention away from the unprotected *rhinophores* and *oral tentacles.* If damaged or removed by predators, the *cerata* are quickly regenerated. The *cerata* function as a respira-

tory surface, and inside each is a glandular digestive lobe. The *cerata* may also contain specialized cnidosacs in which are located undischarged nematocysts taken from their coelenterate prey. (Coelenterates are alternately termed cnidarians.) Defensive mechanisms present in some species include poison glands, prickly bundles of calcareous spicules (sharppointed structures of calcium carbonate), or noxious mucous secretions.

Like the dorid, the aeolid has a pair of *rhinophores;* it also possesses pairs of elongated *oral tentacles* and *propodial tentacles* located at the front of the *foot.* These additional sensory structures aid the aeolid in finding and attacking its coelenterate prey.

The aeolid feeds primarily on hydroid colonies. Its elongated body and long narrow *foot* (5 cm) are adapted to clinging on the erect tree-like hydroid.

Aeolid nudibranchs are seasonally common on the Pacific coast of the United States, and occur both in the rocky intertidal zone and around floats and piers in quiet harbors and bays.

Now color the sea hare. Note that the parapodium (B¹) is an extension of the foot and receives the same color. You may wish to color the algae green or red.

The sea hare, a close relative of the nudibranch, is named for its large, rabbit ear-like *oral tentacles* and its voracious, herbivorous appetite. It is among the largest of the sea slugs; some species reach 40 centimeters in length and may weigh up to 2 kilograms (about 5 pounds).

The sea hare possesses a pair of *rhinophores;* its gill is covered by the *mantle.* The *foot* has two broad wing-like flaps called *parapodia,* which can be folded over its back or flap to create a respiratory current over the gills.

Defensive adaptations of the sea hare include its secretion of a distasteful milky substance, and the glandular ejection of a vivid purple dye. In their *mantles,* sea hares store noxious organic compounds, garnered from algae, that further deter predators.

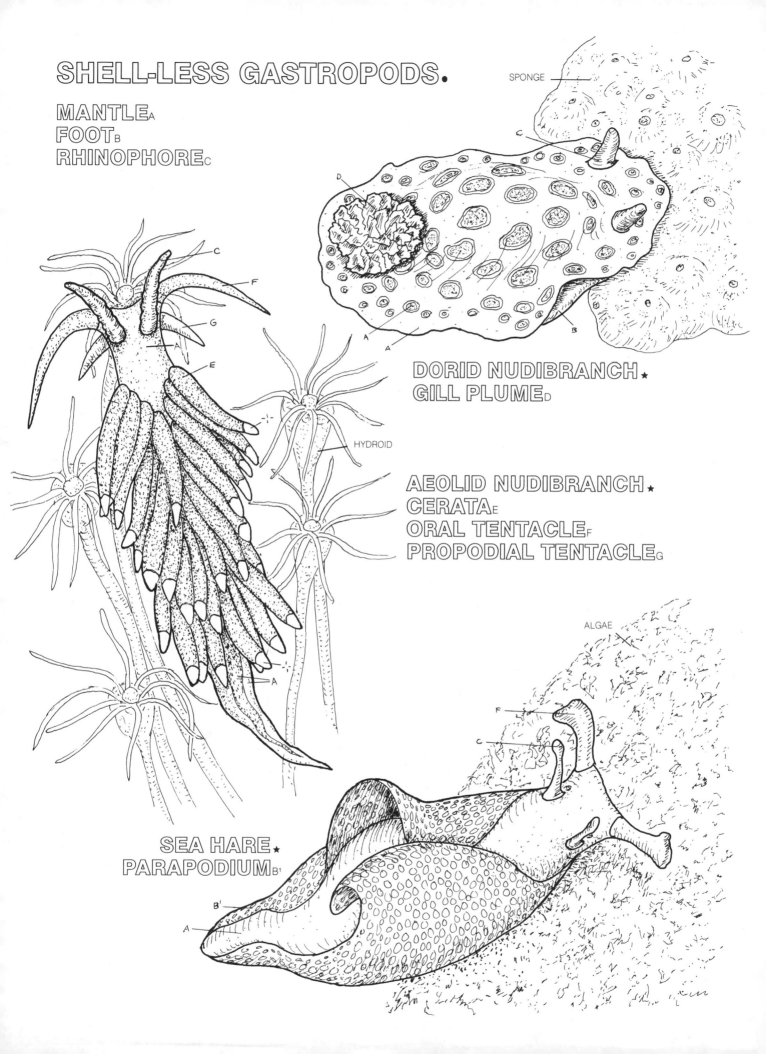

SHELL-LESS GASTROPODS.

MANTLE A
FOOT B
RHINOPHORE C

SPONGE

DORID NUDIBRANCH ★
GILL PLUME D

HYDROID

AEOLID NUDIBRANCH ★
CERATA E
ORAL TENTACLE F
PROPODIAL TENTACLE G

ALGAE

SEA HARE ★
PARAPODIUM B¹

26
MOLLUSCAN DIVERSITY: NAUTILUS

The class Cephalopoda contains the squids, octopuses, nautiluses, and related forms. The cephalopods have a highly developed head with large, well-organized *eyes*.

Begin by coloring the nautilus in its shell and the cut-away shell section.

The chambered nautilus, because of its *shell,* is considered to be the most primitive living cephalopod. The *shell* of the nautilus is partitioned into *chambers* by the *septa,* which are made of the same material. As the animal grows, it creates a new, larger *chamber,* and occupies only the last largest *body chamber* in the *shell.*

Color the illustration of the nautilus removed from its shell. The body of the nautilus receives the same color as the body chamber to indicate the animal's position in its shell.

The animal has a large number of *tentacles* (more than ninety!), topped by the leathery *hood* that is formed by two specially folded *tentacles.* The *eyes* of the nautilus, less well developed than the eyes of other cephalopods, lack a lens and operate somewhat like a pinhole camera.

The *funnel,* located below the *tentacles,* is an evolutionary development of the ancient molluscan foot. The sac-like *body* contains the viscera and the gill chamber (not shown), which opens to the outside via the *funnel.* The tail-like structure protruding from the rear dorsal area is the *siphuncle,* a continuous cord of special secretory tissue which penetrates the chamber walls, or septa, through the *siphuncle openings.*

Color the bottom three drawings depicting the ways in which the nautilus moves. In order to show how the siphuncle and the chambers are used in buoyancy control, the third illustration shows only the nautilus shell. You may wish to color the nautilus the same colors you used in the top drawing.

There are only six known species of nautilus, and most live in the deep waters of the tropics. Consequently, relatively little is known about the biology of these animals. It is known, however, that the nautilus has developed three modes of movement. By day, the nautilus remains on the bottom of the ocean, either resting with its *tentacles* retracted or holding onto the bottom by use of its *tentacles.* Once free of the bottom, the nautilus swims by a kind of jet *propulsion,* like all cephalopod molluscs. Water is forced out of the mantle cavity through the *funnel* by a retraction of the body into the shell and a contraction of the funnel musculature, propelling the nautilus backward.

At night, the nautilus rises to shallower waters by increasing its *buoyancy.* To achieve this, gas is secreted into the *shell chambers* through the *siphuncle,* and water is removed from these *chambers.* This lessens the weight of the nautilus, and it floats upward. Once in shallow waters, the nautilus swims to reefs or rocky areas to feed. Here its *tentacles* are used for movement and to capture slow-moving fish and invertebrate prey. When increasing light signals the approach of day, the gas in the *chambers* is reabsorbed and replaced by water, and the nautilus sinks back into its deep water retreat.

NAUTILUS.

SHELL_A
SEPTUM_{A¹}
CHAMBER_B
 BODY CHAMBER_{B¹}
SIPHUNCLE OPENING_C

ANIMAL_D
HEAD ★
 EYE_E
 TENTACLE_F
 HOOD_G
 FUNNEL_H
BODY_{B²}
SIPHUNCLE_{C¹}

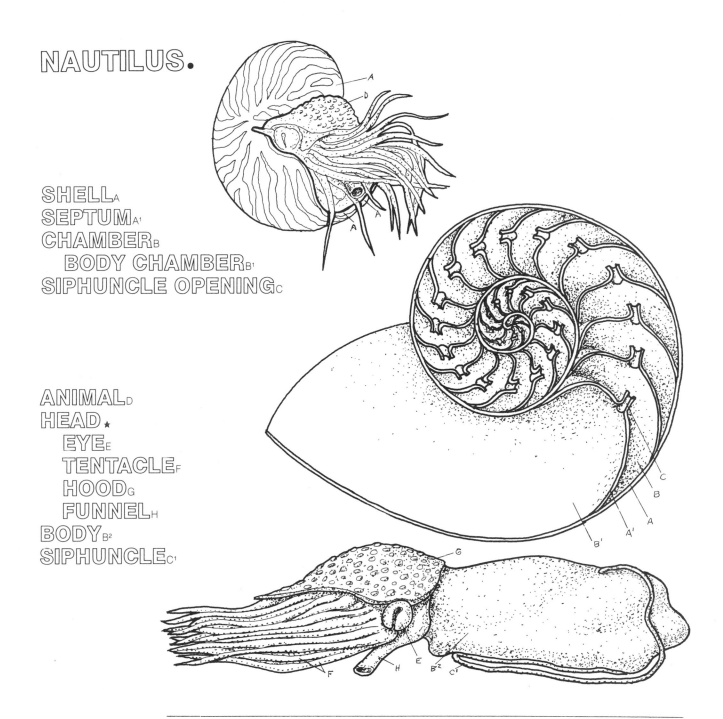

MOVEMENT ★
CRAWL_{F¹}
PROPULSION_{H¹}
BUOYANCY_{C²}

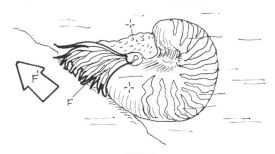

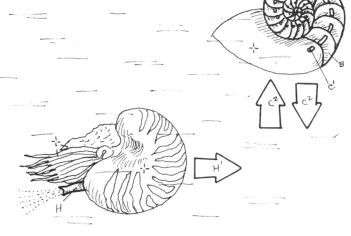

MOLLUSCAN DIVERSITY: SQUID AND OCTOPUS

The squid and the octopus are two highly developed members of the class Cephalopoda. While the chambered nautilus relies for buoyancy on the ancestral molluscan shell, the squid's thin shell is located within the mantle and is useful only for muscle attachment; the octopus has lost its shell entirely.

Begin by coloring the large illustration of the squid.

The squid swims by forcing water out the *funnel,* in a jet propulsion fashion. This, with its tapered, streamlined *body* and the broad triangular *fins* used for stabilization, makes the squid a highly effective swimmer. Normally, the squid swims backward. Squids may also swim forward, by directing the *funnel* posteriorly, and are capable of hovering motionless in the water. Over short distances, squids are among the most rapid moving of all marine organisms. Large squids can attain speeds of 24-32 kilometers (15-20 miles) per hour.

The muscular mantle of the squid and the mantle cavity it houses are strengthened internally by plates of cartilage in the body wall and by the remnant of the shell (called the pen, not shown). The eight *arms* and two *tentacles* can be held motionless in front of the squid's *head* to aid in streamlining. The *funnel* is both moveable and muscular, so that in swimming, the opening (lumen) can be constricted, thereby increasing the pressure of the water forced out of it by the contraction of the mantle cavity.

Color the enlargment of the front view of the squid and the drawing of the beak.

Its swimming ability, coupled with its image-forming *eyes,* gives the squid a tremendous advantage as a predator. It can swim into a school of fish and quickly capture one with its long sucker-tipped *tentacles.* The fish is dispatched with a bite behind the head from the *beak* of the squid — accompanied by an injection of poison. The *beak* is located in the center of the circle of *arms,* protruding from the mouth; it cuts the prey into small pieces that are then carried into the mouth by the radula (not shown).

Located on the *arms* are stalked, adhesive discs, or *suckers* (see circled enlargement), which in some species are reinforced by horny rings or hooks. Contraction of the muscles attached to each *sucker* creates suction when the *suckers* come in contact with something solid. The *tentacles,* twice as long as the *arms,* have *suckers* only on their flattened ends.

Color both illustrations of the octopus. Note that the beak is not shown here. Also color the enlargement of the suckers, showing their shape and relative size.

The octopus does not normally swim about in the water. It will swim, however, if threatened. It swims with its soft bag-like *body* held in the direction of movement, and its *head* and eight *arms* trailing behind. The *funnel* is directed rearward and the octopus moves in typical cephalopod fashion, propelled by forcing a jet of water out the *funnel.* The octopus lacks the streamlining that makes the squid such a successful swimmer; it prefers to remain in contact with a solid structure, pulling itself along using the *suckers* on its *arms.*

In most octopus species there are about 240 *suckers* on each *arm,* usually arranged in double rows. The *suckers* lack the stalk, the horny rims and the hooks possessed by the squid. Octopus *suckers* vary in size from a few millimeters to 7 centimeters in diameter. A *sucker* 2 centimeters in diameter requires a pull of six ounces to break its hold, so one can imagine the strength it would take to break a hold of two thousand suckers!

The octopus generally is a solitary dweller and seeks shelter or a permanent den in a cave or under rocks.

SQUID.

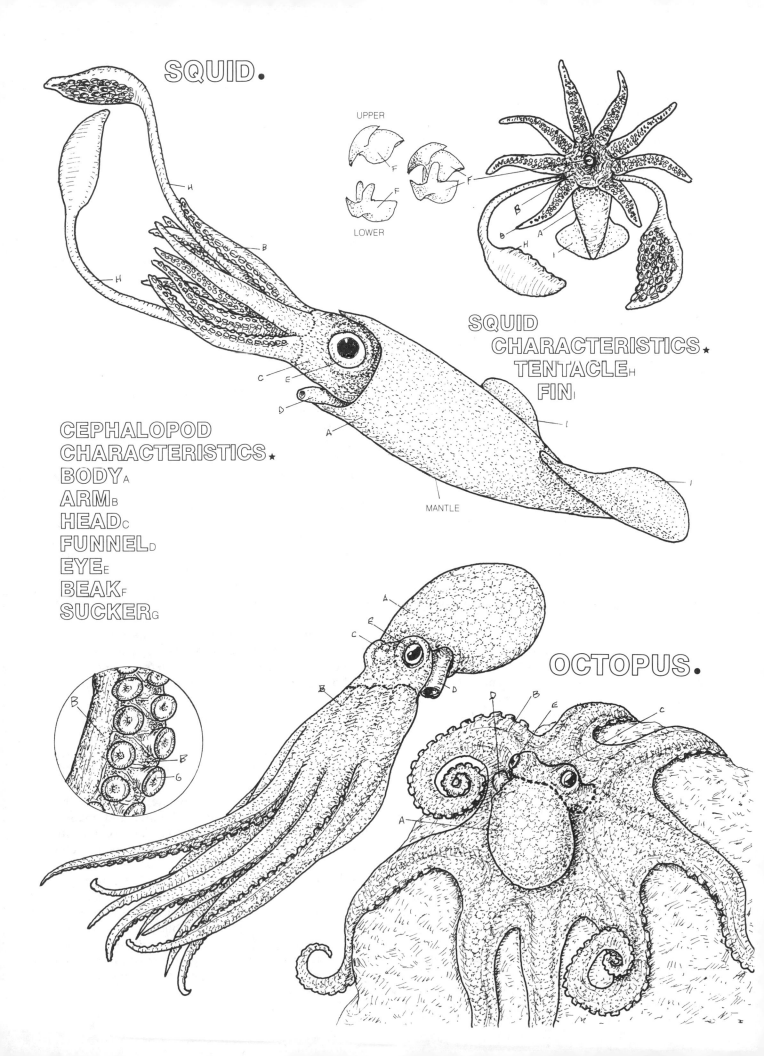

UPPER
F
F
LOWER

SQUID
CHARACTERISTICS ★
TENTACLE H
FIN I

CEPHALOPOD
CHARACTERISTICS ★
BODY A
ARM B
HEAD C
FUNNEL D
EYE E
BEAK F
SUCKER G

MANTLE

OCTOPUS.

CRUSTACEAN DIVERSITY: SMALL CRUSTACEANS

Crustaceans belong to the phylum Arthropoda, which also includes the terrestrial insects (flies, bees, ants, etc), and is the most abundant animal group on earth, both in total number and in number of species.

All arthropods share two characteristics: they are encased by an external skeleton, or exoskeleton, and their bodies and appendages are jointed, or segmented. Crustaceans also have a chemical complex of calcium carbonate within their exoskeletons, giving them a "crusty" texture, hence their name. Five very common small crustaceans are illustrated here.

First color the copepod in the upper right corner of the plate. Antennal movement is indicated by the broken line. The small arrows indicate the direction of water current flow. Color the fine hairs on the mouth parts.

The copepod is among the smallest (2.5-10 mm) and most abundant of crustaceans and is usually a dominant member of marine zooplankton. The most prominent feature of the copepod is its elongated *antennae,* which extend at right angles from the anterior *prosome.* As these *antennae* are flexed backwards in a jerking, sculling motion, they both propel the copepod and generate currents around the copepod's filter-feeding *mouth parts.* Another prominent structure is the single, light sensitive *eyespot.*

The segmented copepod body bends only at the junction of the anterior *prosome* and the posterior *urosome.* The *urosome* terminates in the *telson* with its two extensions, called the *caudal rami* (singular: ramus).

Now color both the acorn and the stalked barnacles. Color over the fine hairs of the cirripeds on both barnacles and on the enlarged illustration. Use a light color so that the filtering hairs show through.

Acorn barnacles (5-50 mm in diameter) attach themselves to solid substrata, such as rock, pier pilings, and the bottom of ships and of whales. The accumulation of acorn barnacles (and other organisms) on the bottom of a ship is called "fouling," and may increase its drag and decrease its speed and fuel efficiency by 20 percent or more.

From within the fortresslike *shell plates,* the barnacle extends three to six pairs of segmented, biramous (two branched) *cirripeds,* which can be held stationary like a plankton net, or swept rhythmically,

catching whatever the water brings.

Stalked barnacles are characterized by fleshy, flexible, muscular *stalks,* which are capable of lengthening (moving the feeding apparatus out into the water column) and contracting (pulling the barnacle closer to the substratum). Stalked barnacles are found on floating organisms and debris, or attached to hard substrata. They range in size from 5-25 millimeters in diameter.

Color the two diagrams of the isopod on the lower right of the plate.

Isopods vary in size from 1 to 275 millimeters. Illustrated here is the rock slater, a relatively large (50 mm) isopod, related to the common terrestrial pill or sow bugs. Prominent in the dorsal view are the large *compound eyes* and *antennae* of the *head* as well as the serial body segments. The anterior-most seven segments, called the *pereon,* are shown in the ventral view together with the corresponding appendages called *pereopods* or walking legs, all of which are very similar in shape and give the isopods their name (iso: same, pods: legs). Also seen in the ventral view is the posterior body region called the *pleon* and its appendages known as *pleopods,* used in swimming and respiration. The final segment is the *telson,* which is flanked by the *uropods.* At the anterior end, as seen in the ventral view, are the *mouth parts* that are used to feed on scavenged material in the intertidal zone.

Now color the amphipod on the lower left.

This crustacean is from the "gammarid" amphipod (amphi: double, pods: legs) group, which includes the familiar beach hoppers. Their structure is somewhat similar to the isopod just colored. Note that the front two pairs of *pereopods* are claw-like; these are called *gnathopods* and are used in feeding and mating. Behind these *gnathopods* are five pairs of walking legs, two pairs facing backward and three pairs facing forward. Behind the *pereon* is the *pleon* with three pairs of *pleopods* used for swimming. The most posterior segments constitute the *urosome* whose appendages *(uropods)* are used in jumping. First, the entire *urosome* is tucked under the body, then suddenly straightened, springing the amphipod into the air. This well-developed jumping ability of sandy-beach amphipods has earned them the common names "beach hopper" and "sand flea."

SMALL CRUSTACEANS.

COMMON
STRUCTURES ★
ANTENNA_A
EYESPOT/
   COMPOUND EYE_B
MOUTH PARTS_C
TELSON_D

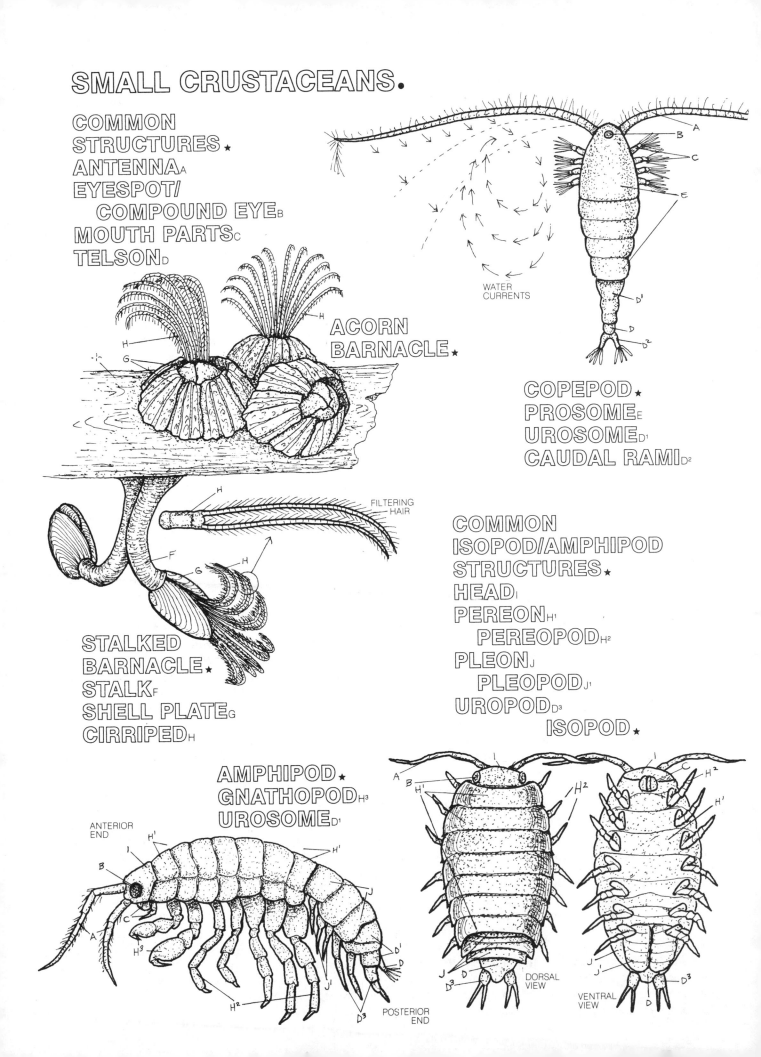

WATER
CURRENTS

ACORN
BARNACLE ★

COPEPOD ★
PROSOME_E
UROSOME_D1
CAUDAL RAMI_D2

FILTERING
HAIR

COMMON
ISOPOD/AMPHIPOD
STRUCTURES ★
HEAD_I
PEREON_H1
   PEREOPOD_H2
PLEON_J
   PLEOPOD_J1
UROPOD_D3

STALKED
BARNACLE ★
STALK_F
SHELL PLATE_G
CIRRIPED_H

ISOPOD ★

AMPHIPOD ★
GNATHOPOD_H3
UROSOME_D1

ANTERIOR
END

POSTERIOR
END

DORSAL
VIEW

VENTRAL
VIEW

29
CRUSTACEAN DIVERSITY: DECAPODS

Most of the larger familiar crustaceans belong to the order Decapoda (deca: ten, poda: legs).

Color each decapod crustacean as it is covered in the text. Begin by coloring the shrimp, for which only one of each pair of appendages is shown.

The long, jointed rear portion of the body of the shrimp is called the *abdomen*. Five of the abdominal segments have appendages known as *swimmerets*, which beat rhythmically and are used in swimming. Located on the last abdominal segment are the *uropods*, a pair of appendages that flank the *telson*: together these three comprise the tail fan of the shrimp.

The large, smooth, saddle-shaped structure at the anterior end of the shrimp is the *carapace*, which covers the head and the thorax. Attached to the thorax are the five pairs of thoracic appendages — the ten legs that give the decapods their name. These appendages are typically used as *walking legs*. The front two or three pair of legs have claws and are called *chelipeds* (claw feet). Protruding anteriorly are the two *maxillipeds* (jaw-feet) that are used for feeding, and the *antennae*. The *eyes* of the shrimp are stalked. Although shrimp can swim quite well, many tend to spend most of their time on or near the ocean bottom.

Now color the lobster, and note the difference in the shape of the right and left chelipeds. Only one of each pair of appendages is shown here, with the exception of the uropods, chelipeds, antennae, and maxillipeds.

The large American lobster is a bottom dweller, found in the colder waters off the Atlantic coast of the United States. The lobster is heavy-bodied compared to the shrimp, with a large *abdomen* and huge *chelipeds*. The two *chelipeds* differ in shape and size and perform two functions. The more massive *cheliped* (called the "crushing claw") is used to break through hard-shelled invertebrate prey, mainly molluscs. The more slender "cutting claw" is lined with sharp teeth and is used for shearing the prey into small pieces.

Although the *abdomen* has only rudimentary *swimmerets* (not shown), a rapid flexure of the abdominal segments, coupled with spreading of the tail fan (*uropods* and *telson*), quickly propels the lobster backward. A series of these contractions will rapidly remove the lobster from immediate danger.

Color the hermit crab which is shown here in its shell, and also illustrated in a "see-through" shell. Do not color the shell in either illustration.

Hermit crabs are found in shallow water throughout the world's oceans. They generally inhabit the abandoned shells of gastropods, rarely leaving the security of this "borrowed" home. The *abdomen* of the hermit is large and its exoskeleton lacks the calcification typical of most crustaceans; thus the *abdomen* is soft and vulnerable when not protected by a shell. The *abdomen* is asymmetrically shaped to fit into the tapering spiral of the interior of a gastropod shell. The *uropods* are small, enabling them to cling to the interior of the shell at its apex. There are no *swimmerets* located on the right side of the coiled *abdomen*, and females have *swimmerets* only on the left side, to which her eggs attach. The crab's two posterior pairs of *walking legs* are small; the hooks on the ends hold the hermit in its shell.

Protruding from the shell are the well-armored *chelipeds*, two pairs of *walking legs*, the stalked compound *eyes*, and the *antennae* and *maxillipeds*. With its shell, the hermit travels over the substratum, scavenging for food. If danger approaches, quick contraction of its abdominal muscles pulls the hermit into the safety of its shell.

Now color both drawings of the sand crab. Note the enlarged antennae and the antennal hairs which are used for filter feeding.

The mole crab, or sand crab, is frequently found on wet, sandy beaches of both the Atlantic and Pacific coasts of the United States. Sand crabs follow the tide up and down the beach and must frequently reburrow in their wave-churned habitat. They take advantage of the abundance of suspended particles carried by the constantly moving water. Their egg-shaped *carapace* is easily pulled into the sand by the rapid digging motions of their *uropods* and fourth pair of *walking legs*. The *abdomen* flexes underneath the *carapace*, and they burrow, facing into the waves, extending their long *antennae* into the moving water. The stout antennal hairs trap suspended particles, which are then removed by wiping the *antennae* across the mouth.

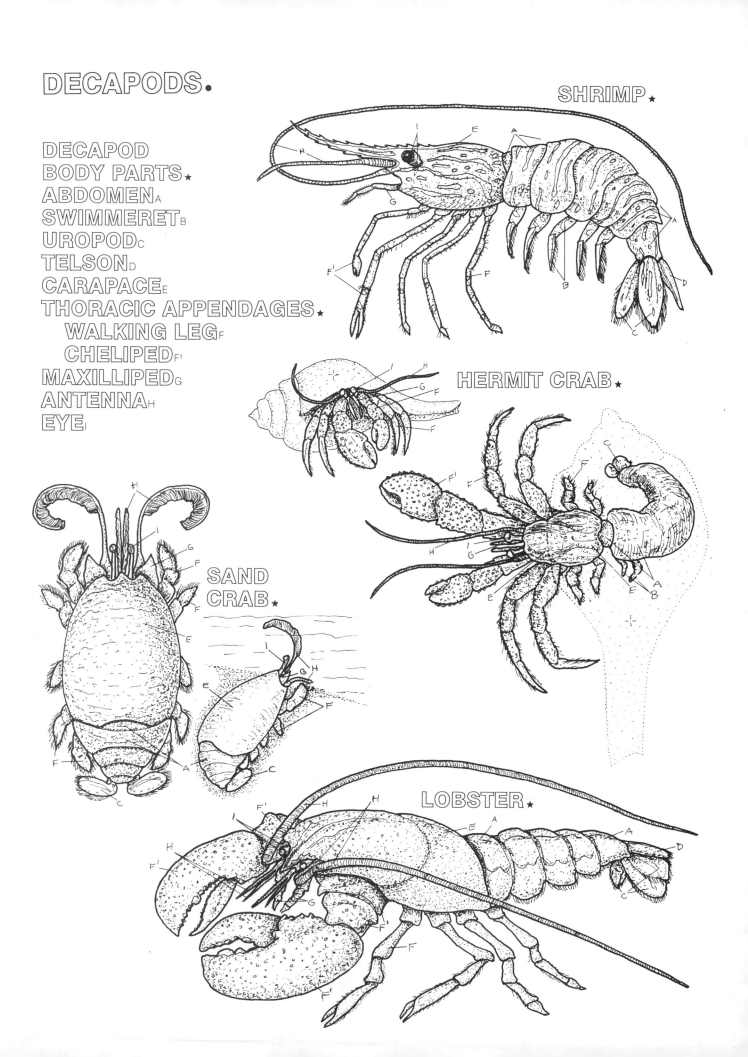

DECAPODS.

SHRIMP ★

DECAPOD
BODY PARTS ★
ABDOMEN A
SWIMMERET B
UROPOD C
TELSON D
CARAPACE E
THORACIC APPENDAGES ★
    WALKING LEG F
    CHELIPED F1
MAXILLIPED G
ANTENNA H
EYE I

HERMIT CRAB ★

SAND
CRAB ★

LOBSTER ★

30
CRUSTACEAN DIVERSITY: CRABS

Color each crab separately as it is discussed in the text, paying particular attention to the differences in morphology. The small drawings show the crabs in their various environments, and need not be colored.

There are approximately 4,500 species of true crabs (section Brachyura), and they are found in a wide variety of marine habitats. All true crabs possess ten thoracic appendages (decapods) and a small, flexed *abdomen*.

The cancer crab has four pairs of *walking legs*, a pair of chelipeds terminating in a moveable finger called the *dactyl*, and a palm-like *manus*. The *abdomen* is visible where it folds behind the shield-like *carapace*, and tucks against the underside of the crab. The *antennae*, mouth parts, and the stalked compound *eyes* are also prominent features of the crab.

Cancer crabs inhabit intertidal and near-shore habitats in the colder and temperate waters of the earth's oceans. Some cancer crabs, such as the Pacific coast market crab, live on sandy bottoms and have relatively thin exoskeletons, allowing fast movement across the substratum. However, most (such as the one pictured here) have a very heavy, thick exoskeleton for protection. Although the weight of the exoskeleton slows the crab's movement, the crab's large chelipeds are quite strong and useful in crushing prey such as snails and bivalves, whose capture does not require speed. The claws of the male crab increase in size when sexual maturity is attained.

The shore crab inhabits the rocky intertidal zone, where it is a very agile climber and rapid crawler. The *walking legs* of the shore crab are relatively short, and the last segment is pointed for clinging tightly to the wave-swept substratum. Shore crabs are among the most conspicuous of the rocky intertidal crabs, as they live higher in the intertidal zone and may be active during daylight hours and low tides. Many shore crabs are herbivores with chelipeds specially modified for cropping the short algal turf growing on the rocks. The tips of the *dactyl* and *manus* form a spoon shape to enable the crab to bring algae and detritus to its mouth.

The blue crab of the Atlantic and Gulf of Mexico belongs to the group of swimming crabs. The two distal leg segments on the last pair of *walking legs* are flattened into paddles. The rear pair of *legs*, which are oriented vertically in non-swimming crabs, are pivoted to a near-horizontal position in the blue crab. These *legs* are flexible and can be rotated over the *carapace* in a sculling action that gives the crab both lift and propulsion. Over short distances, blue crabs can swim at speeds up to one meter per second.

As a predator, the blue crab burrows into soft sediment, with only its *eyes* and *antennae* exposed, where it waits for prey. The blue crab swiftly attacks with its sharp-toothed chelipeds and can pluck a small fish right out of the water.

The name "box crab" refers to the squarish *carapace* that completely overhangs the *walking legs* when this crab is at rest. The hairy, cockscomblike ridges on the box crab's chelipeds help it to burrow in the soft sandy bottom. The chelipeds are held tightly against the body, and the incoming respiratory current is filtered through the hairy ridges. The curious ornamentation on the box crab's chelipeds has given rise to another common name for the crab: the rooster crab.

An even more fitting name for the box crab is the surgeon crab. The crab's chelipeds are asymmetrical; the right cheliped is equipped with a pronounced tooth on the *dactyl* that fits between two teeth on the opposing *manus*. This claw is inserted into the aperture of a gastropod, and the edge of the shell is crushed between these teeth. A large box crab will daintily snip along the gastropod's shell until the soft body of the prey is exposed.

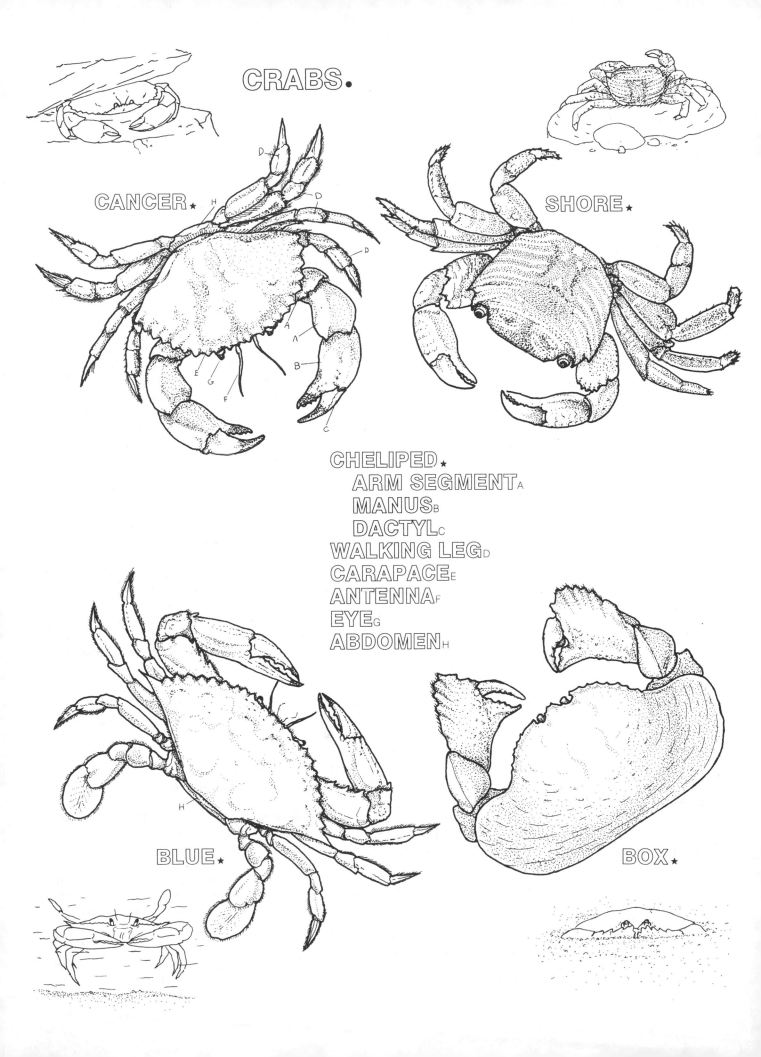

CRABS.

CANCER ⭐

SHORE ⭐

CHELIPED⭐
ARM SEGMENT A
MANUS B
DACTYL C
WALKING LEG D
CARAPACE E
ANTENNA F
EYE G
ABDOMEN H

BLUE ⭐

BOX.

31

ECHINODERM DIVERSITY: SEA STAR STRUCTURE

The five-armed sea star is a widely recognized marine animal, belonging to the echinoderms ("spiny-skins," phylum Echinodermata).

Color the two views of the sea star. Note that the upper (aboral) surface is divided into a central disc and five rays.

The body plan of echinoderms shows a pattern of radial symmetry. That is, the basic and more or less equal units are arranged in a circle around a central *disc,* rather than astride a single midline, as in bilaterally symmetrical animals (like us). Many sea stars and other echinoderms have five of these similar units, giving them pentamerous *(penta:* five, *mer:* part*)* radial symmetry.

The *mouth* of a sea star is located on its oral surface, and the opposite side is called the aboral surface. The aboral side of the sea star has a highly textured, bumpy surface, sometimes with spines. These bumps and spines are parts of the endoskeleton, which is covered by a layer of epidermis. The endoskeleton consists of individual *ossicles* which fit together closely in some species, more loosely in others. This tightness or looseness of fit determines the degree to which an echinoderm's body is stiff and tough or, on the other hand, soft and flexible.

The oral surface of the animal bears the centrally located *mouth.* The oral surface has furrows running from the *mouth* along the length of each *ray.* These furrows are called *ambulacral grooves* and are bordered by one or more rows of *tube feet.* The spiny endoskeleton is also visible on the sea star's oral surface.

Now color the sectional cutaway illustration of the sea star. Note how the canals are connected in series from the madreporite to the ampullae of the tube feet. Also color the enlargements of the tube foot mechanism.

The water vascular system of the sea star runs from the *madreporite* through the *stone canal* to the *ring canal,* from there through the five *radial canals,* and finally to individual *tube feet.* This unique system operates through a combination of muscles and hydraulic pressure (internally modified sea water).

Each *tube foot* consists of a hollow, muscular structure attached to a balloonlike fluid reservoir called an *ampulla.* The elastic ampullar surface is covered with a meshwork of muscle fibers. When these muscles contract, the *ampulla* is deflated and fluid is forced into the *tube foot.* This stretches the *tube foot* musculature (not shown) and extends the *tube foot* outward beyond the *ambulacral groove.* Contraction of the *tube foot* muscles forces the fluid back into the *ampulla* and stretches the ampullar musculature. Many sea stars have muscle fibers attached to the bottom of the *foot.* When the bottom of the *tube foot* is pressed against a solid surface, contraction of these muscle fibers creates a vacuum, allowing the *tube foot* to operate as a suction cup.

The water vascular system, involving hundreds of *tube feet,* is the basis for the locomotor ability of the sea star. Each *tube foot* is under fine nervous control: not only can it be extended, but it can also be moved through a 360 degree arc by local contraction and relaxation of the musculature, playing against the hydraulic pressure from the *ampulla.* The sea star is capable of mostly quite slow, but very precise and well-coordinated movements on nearly any solid horizontal or vertical substratum. This system would not be effective, however, without some rigid structural support for the muscles. This is provided by the interconnecting *ossicles* of the endoskeleton.

The drawing at the top right shows a sea star prying open a mussel. Color this sea star, but not the mussel.

Using its *tube feet,* the sea star can capture live prey and manipulate it into its *mouth.* It can weaken the strong adductor muscles of a bivalve mollusc (like a clam or mussel) by exerting a strong and continuous pulling with its suckered *tube feet.* The sea star needs to force open the bivalve shells only a few millimeters or so; it then extrudes its stomach into the opening and digests the bivalve externally.

Sea stars are usually carnivorous predators, and live on all types of ocean bottoms. Most sea stars have five *rays* and range from 10 to 25 centimeters in diameter. Some species may be much larger and have more than five *rays:* the sunflower star of the Pacific coast of the United States has twenty-six or more *rays* and often reaches one meter in diameter (see Plate 8).

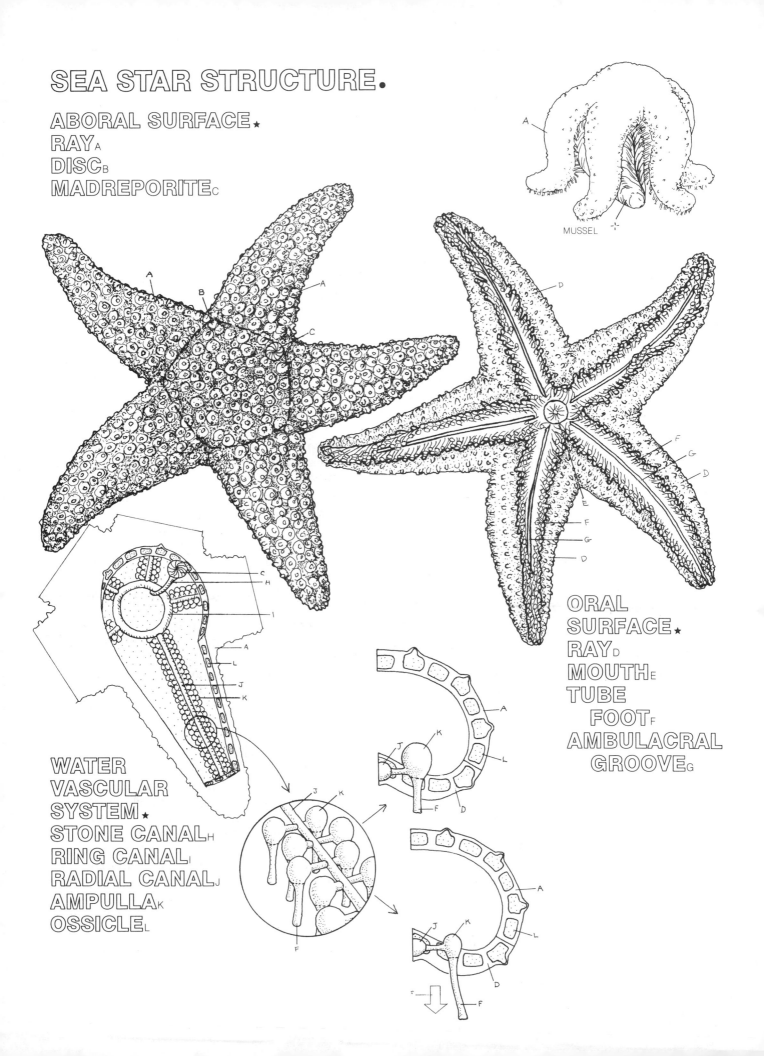

SEA STAR STRUCTURE.

ABORAL SURFACE★
RAY A
DISC B
MADREPORITE C

MUSSEL

ORAL
SURFACE★
RAY D
MOUTH E
TUBE
FOOT F
AMBULACRAL
GROOVE G

WATER
VASCULAR
SYSTEM★
STONE CANAL H
RING CANAL I
RADIAL CANAL J
AMPULLA K
OSSICLE L

32

ECHINODERM DIVERSITY: OPHIUROIDS AND CRINOIDS

The ophiuroids and crinoids are two other classes, which, along with the sea star, are members of the phylum Echinodermata.

Begin by coloring the small sea star on the right side of the plate and the middle cutaway side view showing the location of its aboral surface, oral surface, and mouth. Color the cutaway illustrations of the ophiuroid (top) and the crinoid (bottom) showing the location of the above structures in these two organisms.

Finally, color the illustration of the ophiuroid, the enlargment of the arm section, and the drawing at the upper right, showing ophiuroid movement and autotomy.

Like the sea star, the ophiuroid, or brittle star (also called serpent star) has a five-rayed star pattern. The *mouth* is located on the *oral surface,* and the opposite side is called the *aboral surface.* Unlike the sea star, whose rays and disc merge together, the rays, or *arms,* of the brittle star are quite distinct from one another. These *arms* are thin and usually spiny, each consisting of a row of large articulating structures, much like a vertebral column. These structures are skeletal *ossicles* which give the *arm* an undulating, snake-like *movement.*

The muscles that link the articulating *ossicles* are capable of violent contraction, a defense mechanism the brittle star will use if it is captured or trapped. The contraction may cause the trapped *arm* to sever, in a phenomenon known as *autotomy* (self-cut), and allows the ophiuroid to escape, leaving part of the *arm* behind. The common name "brittle star" is derived from this ability to cast off an *arm* if handled. As in sea stars, severed *arms* are later regenerated.

Many ophiuroid species are omnivorous scavengers. They use their tube feet to catch suspended food material in the water, or to pick food off the bottom. The *arms* are used in locomotion: two *arms* are held forward, two out to the side, and one to the back. The *disc* is held up off the substratum and the laterally positioned *arms* move in a rowing motion that propels the animal along in short leaps. The *spines* on the *arms* help to keep the animal from slipping on the

substratum. Ophiuroids are the most rapidly moving of all the echinoderms.

Ophiuroids comprise over 2,000 species. Some smaller species that can withstand its rigors inhabit the intertidal shoreline among the rocks, but most species live in subtidal habitats, from shallow water to the deepest ocean bottom, or in coral reefs. Most ophiuroids are small, with a *disc* diameter of 1 to 3 centimeters and *arms* from 5 to 6 centimeters in length; a few species have long, tapering arms that reach a length of 15 to 25 centimeters.

Color the crinoid and the enlargement of one of its arms which shows the ambulacral groove and the tube feet.

The crinoids are the only group of echinoderms having a *mouth* that faces upward. Considered by many to be the oldest living class, crinoids include 5,000 identified fossil forms but only 620 living species. Of these, some 80 species are called sea lilies, and remain permanently attached to a substratum by a long stalk. The sea lilies (not shown) live at depths of 100 meters or more.

Other crinoid species (such as the one shown) move or swim about in the tropical intertidal zones and also at great depths. These species cling to various surfaces with *cirri* that sprout from the cuplike *calyx*. The *mouth* and digestive tract are also located in the *calyx*. Also coming from the *calyx* are the long, slender *arms,* from a minimum of ten to as many as two hundred, in some species. From each *arm* rows of *pinnules* extend laterally, giving the appearance of a feather, from which the group's common name derives: feather star. An *ambulacral groove* extends along the length of each *arm,* and laterally into the *pinnules*. These *ambulacral grooves* are flanked by tentacle-like *tube feet* with slender, mucus-secreting papillae along their length. Crinoids are filter feeders; by holding their *arms* upward, plankton and other suspended materials are trapped on the *tube feet.* Food is transferred to the ciliated *ambulacral grooves,* which then carry it bound in mucus to the *mouth.*

OPHIUROIDS AND CRINOIDS.

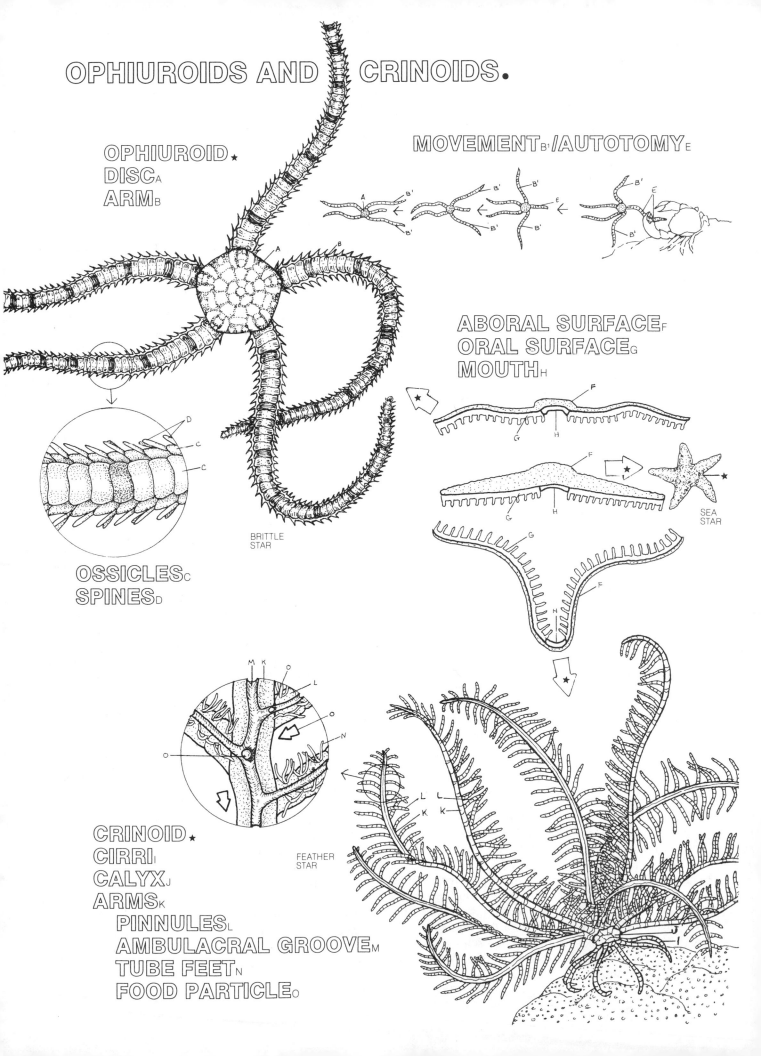

OPHIUROID★
DISC A
ARM B

MOVEMENT B'/AUTOTOMY E

OSSICLES C
SPINES D

BRITTLE STAR

ABORAL SURFACE F
ORAL SURFACE G
MOUTH H

SEA STAR

FEATHER STAR

CRINOID★
CIRRI I
CALYX J
ARMS K
PINNULES L
AMBULACRAL GROOVE M
TUBE FEET N
FOOD PARTICLE O

33
ECHINODERM DIVERSITY: ECHINOIDS AND SEA CUCUMBERS

Unlike other echinoderm classes, echinoids and sea cucumbers (holothuroids) lack arms and have bodies that are stretched along the oral/aboral axis.

Begin by coloring the sea urchin skeleton. Note that the alternating ambulacral (A) and interambulacral (B) sections of the skeleton are arranged like sections of an orange. Use light colors for these sections. Then, color the enlargements of the individual ossicle and the tubercle, including the spines and other appendages.

Color the three drawings of the sand dollar, and note how flat the sand dollar is in the side view, and how the petals (A⁴) are positioned on the aboral surface.

The echinoids include the sea urchins and sand dollars. The individual *ossicles,* or skeletal plates, of the echinoid are tightly sutured or joined, producing a rigid skeleton or "test."

The sea urchin is round or oval in shape, and the individual *ossicles* are organized in ten longitudinal rows, running from the oral to the aboral pole. Five alternating rows are called *ambulacral* plates, and the other five alternating rows are the *interambulacral* plates. The urchin's stalked *tube feet* are located on the *ambulacral* areas; its long, moveable *spines* and defensive *pedicellariae* are attached to both *ambulacral* and *interambulacral* areas.

Sea urchins live primarily on firm substrata. Their *mouth* and oral surface are kept toward the substratum; the *jaws* protruding from the *mouth* scrape and chew off algae and attached organisms. Sea urchins can consume a considerable biomass, and where they occur in large numbers they have a devastating impact on the algae (see Plate 96).

Sea urchins move both by means of their long, stalked *tube feet,* with powerful suckers at the tips, and by means of their *spines.* Each *tube foot* emerges through a pair of holes *(pore pairs)* in the ossicle, as shown in the upper illustration, and moves multidirectionally. The *spines* come in two or more sizes and pivot freely on *tubercles,* or "bosses," on the *ossicle.* The urchin coordinates the *tube feet* and *spines* and walks and pulls itself along. Some urchins with very long *spines* walk quite briskly, using only the *spines* for locomotion.

The jaw-bearing *pedicellariae* protrude from the *ossicles* and are used for defense and cleaning. The *ampullae* and the other structures of the water vascular system are beneath the *ossicles.*

Sand dollars move solely on their small, short spines, which are visible as hairlike structures on the *oral surface* surrounding the *mouth.* The small tube feet (not shown) are found among the spines and assist in gathering food. These animals live on or in soft sediments and are primarily deposit feeders. The *petals* on their *aboral surface* relate to the five *ambulacral* areas of the sea urchin. These *petals* contain special respiratory tube feet (not shown).

Color the illustrations of the two types of sea cucumbers as well as the enlargement of the ossicle. Note the sea cucumber on the right lacks ambulacral/interambulacral areas.

Most sea cucumbers have a thick, leathery body within which are scattered the unattached tiny *ossicles.* This gives the sea cucumber a wormlike flexibility with a potential for burrowing (in fact, many species are burrowers). The elongated body of the sea cucumber creates a "head" area, and "tail," and dorsal and ventral surfaces. Sea cucumbers have a prominent *anus,* and breathe through special respiratory "trees" (not shown) that branch internally from the *anus.* Respiratory water is both brought in and expelled through the *anus* by muscular contraction.

The sea cucumber illustrated on the right is a surface-dwelling deposit feeder. It uses its *oral tube feet* to pick up food from the substratum. The *tube feet* located on the well-developed "sole" are used for movement along the substratum. This cucumber can grow up to 40 centimeters in length.

The sea cucumber on the left is a filter feeder and lives among rocks. This cucumber has five rows of *tube feet* for movement. The *oral tube feet* are highly branched and coated with mucus, which acts to trap suspended plankton and detritus when the *tube feet* are spread open. To feed, the cucumber moves each *foot* into its *mouth* in a rhythmic, systematic fashion, and removes the trapped food particles. Sea cucumbers are found on all types of substrata, from the intertidal zone to the deepest ocean depths.

ECHINOIDS AND SEA CUCUMBERS.

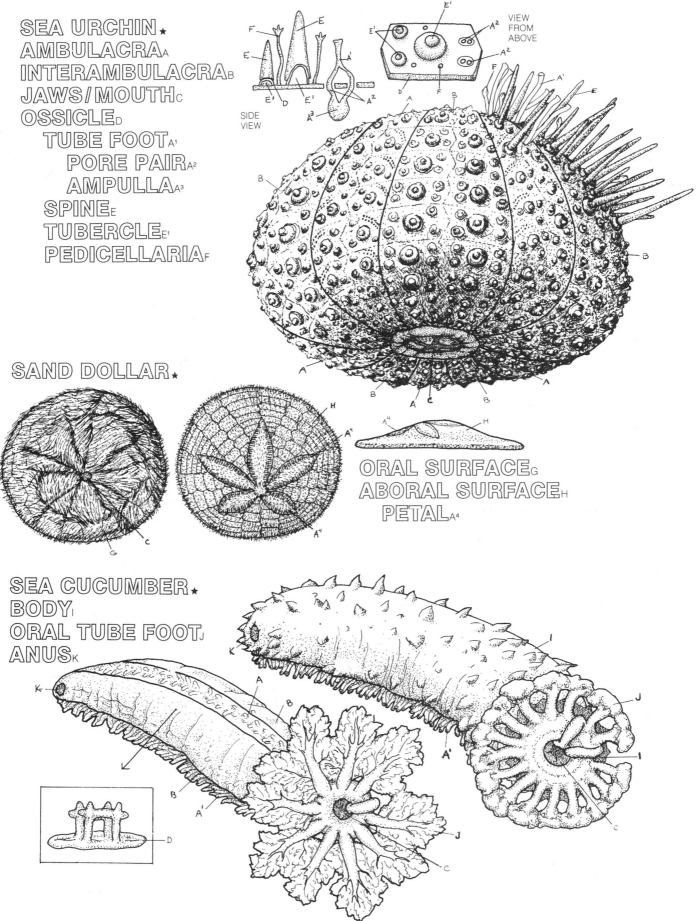

SEA URCHIN ★
AMBULACRA A
INTERAMBULACRA B
JAWS/MOUTH C
OSSICLE D
TUBE FOOT A¹
PORE PAIR A²
AMPULLA A³
SPINE E
TUBERCLE E¹
PEDICELLARIA F

VIEW FROM ABOVE

SIDE VIEW

SAND DOLLAR ★

ORAL SURFACE G
ABORAL SURFACE H
PETAL A⁴

SEA CUCUMBER ★
BODY I
ORAL TUBE FOOT J
ANUS K

34
MARINE PROTOCHORDATES: SEA SQUIRT AND LARVACEAN

The protochordates (beginning chordates) are a group of marine animals that have several characteristics that link them to the higher chordate animals (including fish and mammals). The most successful protochordates are the tunicates (urochordates), comprising 1,300 species of sessile and planktonic forms. Within the urochordates, the largest number of animals belong to the group known as sea squirts.

Begin by coloring the solitary sea squirt attached to the upper portion of the pier piling under water on the left. Use a light color for the tunic. Next, color the long sectional side view and the cross-sectional view of the upper part of the solitary sea squirt on the upper right. The directional arrows indicate water flow, and the stippled arrows indicate movement of food particles; these are not to be colored.

Sea squirt species occur both as solitary individuals and in closely packed colonies. The first sea squirt illustrated here is a medium-sized (7.5 cm) solitary tunicate. The name "tunicate" is derived from the outer covering or *tunic* of the animal, which is thin and transparent in some species and quite thick and leathery in others. This *tunic* contains cellulose (rare in the animal kingdom) and is secreted by the ectodermal tissue of the *body wall*.

Within the *body wall* are circular and longitudinal muscles, which, when contracted, cause water to be squeezed out the *siphon* in a thin stream (hence the name "sea squirt").

The sea squirt is a filter feeder with a simple anatomical organization. Water enters the large *buccal* (oral) *siphon* and flows into the *pharynx*. The *pharynx* is perforated by numerous ciliated *gill slits*. The cilia beat rhythmically and create a water current. Water flows through the *gill slits* into the *atrium*, then exits out the *atrial siphon*. The *endostyle* organ secretes a continuous supply of mucus, which is carried laterally by the ciliated surface of the *pharynx* to form a *mucus sheet*. Food particles are trapped in the *mucus sheet*, and the *sheet* is then taken up by the *dorsal lamina* and shunted to the *stomach*, which opens at the bottom of the *pharynx*. The *anus* and *gonads* both empty into the *atrium*; their products are re-

moved by the excurrent water flow out the *atrial siphon*.

Now color the cluster of sea squirts joined by their tunics on the middle of the piling.

Some species of sea squirts bud asexually, at the base, to form groupings of individuals called social sea squirts. Illustrated here are the medium-size (4–5 cm) sea grapes, commonly found on pier pilings and harbor floats on both the Atlantic and Pacific coasts of the United States.

Now color the compound sea squirts attached to the bottom of the piling. Then color the enlarged cross section of the compound squirt. The directional arrows indicate water flow and should be left uncolored.

A more complex internal structure is seen in the compound tunicates. These tunicates form thin (6-mm thick) sheets over the surface of the rocks, boat bottoms, and almost any solid surface in relatively clean, quiet waters. Concentric groupings of seven or more individual tunicates produce the floral-like pattern seen on these tissue sheets. Each individual has its own *buccal siphon* and *pharynx*, but shares a common *atrium* and *atrial siphon* with its mates.

Now color the larvacean urochordate. The illustration shows this tiny (2–3 mm) animal in its complex mucus house, which is actually transparent and shows only as an outline in this drawing.

The larvacean tunicates retain the characteristics of a larva (not shown) throughout adult life. A larvacean tunicate secretes a *house* of stiff mucus, completely surrounding its *body*; the *house* may be simple or complex, depending on the species.

The larvacean filter feeds, creating a water current by beating its muscular *tail*. Water enters through a coarse-meshed *incurrent filter* that traps large suspended particles. The current then flows through an internal *fine mesh filter* and out the rear *exit valve*. When the *incurrent filter* becomes clogged with debris, the *larvacean* exits out an *escape hatch*. Once free, the animal secretes a new house; this may happen as often as every few hours.

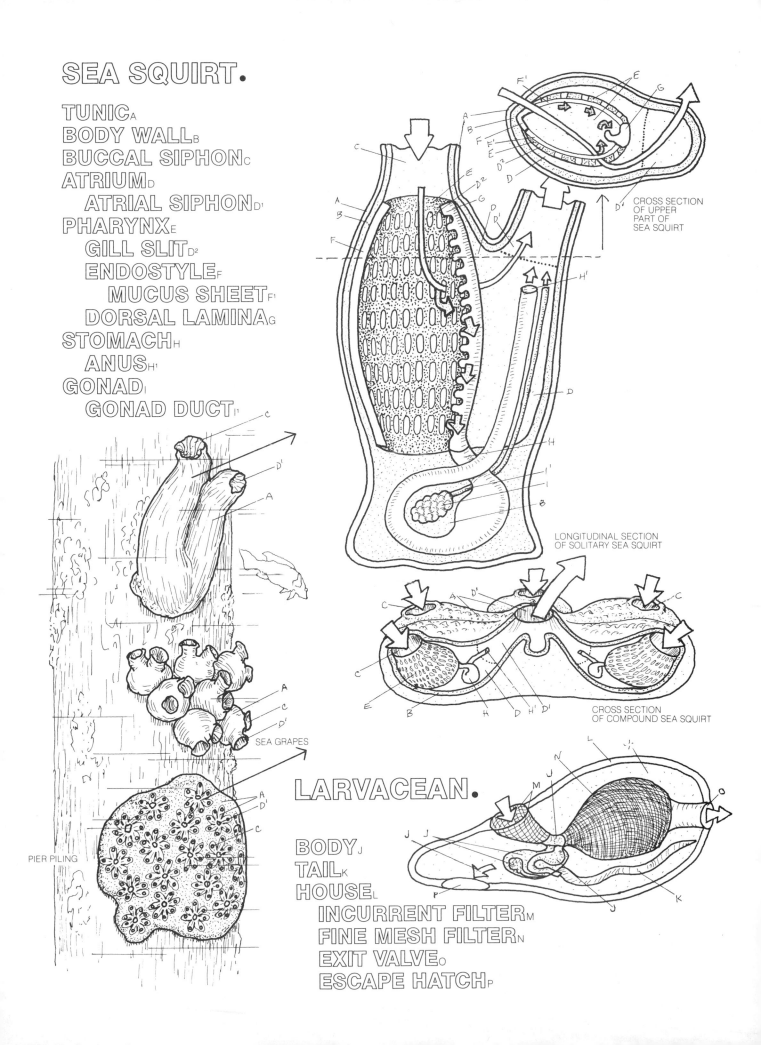

SEA SQUIRT.

TUNIC_A
BODY WALL_B
BUCCAL SIPHON_C
ATRIUM_D
 ATRIAL SIPHON_{D^1}
PHARYNX_E
 GILL SLIT_{D^2}
 ENDOSTYLE_F
  MUCUS SHEET_{F^1}
 DORSAL LAMINA_G
STOMACH_H
 ANUS_{H^1}
GONAD_I
 GONAD DUCT_{I^1}

CROSS SECTION
OF UPPER
PART OF
SEA SQUIRT

LONGITUDINAL SECTION
OF SOLITARY SEA SQUIRT

CROSS SECTION
OF COMPOUND SEA SQUIRT

SEA GRAPES

PIER PILING

LARVACEAN.

BODY_J
TAIL_K
HOUSE_L
 INCURRENT FILTER_M
 FINE MESH FILTER_N
 EXIT VALVE_O
 ESCAPE HATCH_P

35

To appreciate the diversity of marine bony fish, one needs first to investigate the features that are common to most fish. The grouper, or sea bass, pictured here, is considered to be unspecialized, and has a basic fish morphology. The body forms, jaws, and fin shapes of other fish differ to meet the specific demands of their particular habitats.

Begin by coloring the body of the sea bass in the two large drawings. (Do not color other parts of the sea bass now.) Then color the various body forms on the lower left of the plate.

The shape of the sea bass is fusiform—a streamlined shape offering the least resistance to movement through water and allowing the fish to develop a powerful forward thrust with the sweep of the tail or *caudal fin.*

The sea bass *body* tapers to a point at the *mouth*; this enhances its swimming efficiency. Compare this shape to the other *body forms* shown. Fish bodies may be compressed (flattened laterally), depressed (flattened dorsoventrally), elongated, or shortened. Although fish with these other *body* forms are not fast swimmers, they have excellent maneuverability in the water.

Now color the fins in the two illustrations of the sea bass; then color the caudal fins in the lower illustrations and compare the various shapes.

The *caudal fin* usually provides the main thrust used in swimming. The size and shape of this *fin* is an indicator of the fish's ability to move through the water. The sea bass has a rounded *caudal fin* that is soft and flexible, but it also has considerable surface area. This *fin* gives effective acceleration and maneuvering, but is inefficient for prolonged, continuous swimming. A forked *caudal fin* produces less drag, and is efficient for more rapid swimming. Long-distance, continuous swimmers, like the tuna, have lunate *caudal fins,* which are rigid for high propulsive efficiency but render the fish less able to maneuver.

Located on the midline of the fish are two unpaired fins, the *dorsal* and *anal fins.* These *fins* are used to stabilize the fish in the water and to lessen its tendency to pitch, especially while swimming slowly. They are also useful in preventing the fish from roll-

ing over while turning at high speeds. The *pectoral fins* are located on the sides of the fish, behind the opening of the *gill* cavity, and the *pelvic fins* are located ventrally, in front of the *anal fin.* These paired *fins* are also used as stabilizers, but assist in turning and stopping as well.

Fins are supported by fin rays of two types: bony, pointed spines; and soft rays, which are jointed. The *dorsal, anal,* and *pelvic fins* have both spines and soft rays; the remaining fins consist of soft rays only.

Color the mouth/jaws of the sea bass as well as the illustrations of different types of jaws, comparing them to that of the sea bass. Then color all other parts of the sea bass.

In front of each *pectoral fin* is a large bony flap called the *operculum,* which covers the gill cavity that opens just behind it. Running the length of the fish, from the *operculum* to the base of the *caudal fin,* is the *lateral line.* The *line* consists of a series of very small canals that open to the surface and contain pressure-sensitive receptors. When the fish encounters movement in the water, such as the bow wave of an approaching fish, the water pressure pushes against the fish, entering the *lateral line* canals and triggering the pressure receptors. This *lateral line,* called the ''sense of distant touch,'' is extremely sensitive and allows the fish to move in turbid water by ''feeling'' its way around obstacles, even when vision is greatly impaired.

Other prominent external morphological characteristics are the *eyes* (fish have fair vision), *nostrils,* and the bones of the *jaw.* Generally, fish have two *nostrils* on each side, which open to an olfactory pit and are used for scent, not for respiration.

The shape, size, and position of the *mouth* vary considerably in different fish, and are related to the type of feeding. The sea bass has a large *mouth* and is a generalist carnivore, feeding on a wide range of prey. Inside the stout *jaws* of the sea bass is the folded opening of the *esophagus,* which leads to the gut. Adjacent to and in front of the *esophagus* are the *gills,* whose arched gill bars support the gill rakers, sometimes used in feeding, and the gill filaments, which provide the respiratory surface (see Plate 40).

FISH MORPHOLOGY.

BODY A
FINS ★
   DORSAL B
   CAUDAL C
   ANAL D
   PELVIC (2) E
   PECTORAL (2) F
EYE G
JAWS / MOUTH H
OPERCULUM I
LATERAL LINE J
NOSTRIL K
ESOPHAGUS L
GILL M

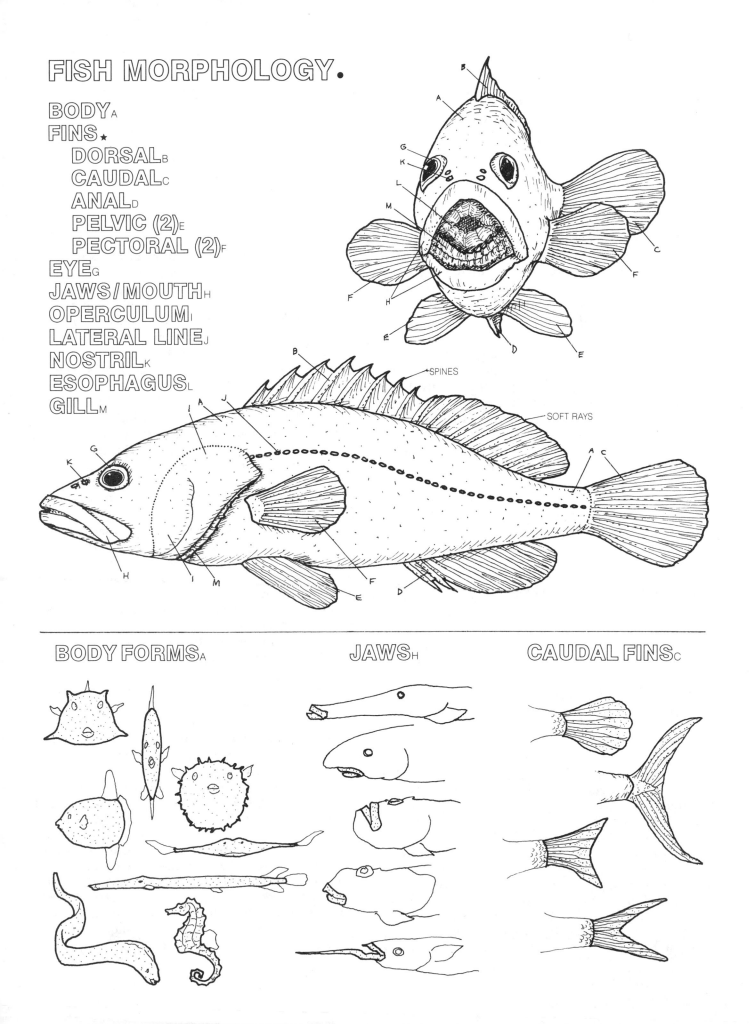

SPINES

SOFT RAYS

BODY FORMS A

JAWS H

CAUDAL FINS C

36
BONY FISH DIVERSITY: PELAGIC FISHES

In the marine environment, fish live either in the water column, or on the ocean bottom. The body structure of the fish is specially adapted to life in one or the other of these environments. Fish that live in the sunlit, open waters, constantly free of the ocean bottom, are called *pelagic fishes*.

All of the fish illustrated here have a similar coloration pattern: dark on the top half, light colored on the bottom. It is suggested that the body of each fish be colored a dark gray-blue on the top half, and a silver or white on the bottom. Structures B through H may be colored the same colors that you used on the previous plate on fish morphology. Color each fish separately as it is discussed in the text.

The flying fish has the capability of becoming airborne, gliding just above the water's surface. If pursued from below by a predator, the flying fish breaks the surface of the water at speeds up to 40 miles per hour. The enlarged *pectoral fins* are stretched out at right angles to the body, and act as a gliding "wing." The *pelvic fins* are similarly enlarged to increase the gliding surface. When the flying fish begins to slow down, it alights on the water, tail first, and the enlarged lower lobe of the *caudal fin* rapidly sculls the surface at a rate of 50 beats per second. The fish picks up speed and it again becomes airborne. Flying fish are found in the warm waters of the Atlantic and Pacific oceans. The largest species reaches a length of 46 centimeters, and is fished commercially off the southern California coast.

The northern herring is a very important small fish feeding on zooplankton, especially copepods. The northern herring occurs on both sides of the Atlantic; a subspecies lives in the Pacific Ocean. The herring forms tremendous schools (shoals) comprising billions of fish. These schools migrate to shallow breeding grounds, where, on the Atlantic grounds alone, two to three million tons of herring are caught yearly. Young herring are canned and sold as sardines; the older fish may reach a length of 0.3 meters, and are either canned or used for oil. Ownership of fishing rights to herring breeding grounds is highly contested, and several European contries have come close to war over this issue.

The swordfish is found in tropical and warm temperate seas worldwide. It is a continuous swimmer, following schools of mackerel, herring, and sardine. The swordfish swims through a school of fish, thrashing its upper *jaw,* or *bill,* and stuns the fish, which it then eats. The swordfish's *bill* may be as large as one third of the fish's entire body length, which averages 1.8 to 3.6 meters. Some swordfish as large as 6 meters have been taken.

The streamlined tapering of the swordfish makes it an excellent swimmer. The swordfish lacks pelvic fins, and the long, low-slung *pectoral fins* are kept folded against the body during rapid swimming. The *dorsal fin* is tall and remains permanently erect. The rigid *caudal fin* is lunate, or crescent shaped, for maximum swimming efficiency; the *caudal* area is strongly reinforced by a bony *keel.*

The sunfish is highly compressed (flattened laterally), and is a slow swimmer. This fish grows to be quite large (3–4 meters) and is the heaviest of the bony fish at 2,000 kilograms (4,400 pounds). The sunfish is usually seen on the surface of the water "sunning" itself on its side, slowly flapping its small *pectoral fins.* Recent studies indicate, however, that the sunfish normally lives quite deep in the ocean's waters, and that those fish seen on the surface are actually quite abnormal.

The sunfish lacks pelvic fins, and the *anal* and *dorsal fins* are very large and set well back on the *body,* where they provide the swimming thrust. The *caudal fin* exists as a narrow band, but is not effective in swimming. The sunfish feeds on jellyfish and other small planktonic forms.

The albacore is a small tuna that averages 4.5 kilograms (10 pounds) and is highly prized as a sport and commercial fish. Albacore schools are found in the Atlantic and Pacific, where they range into temperate waters and spawn near the equator. Like the swordfish, the albacore is a continuous swimmer, with a rigid lunate *caudal fin* and reinforcing *keel.* Besides the *anal* and *dorsal fins,* a series of smaller "finlets" are present and add to the hydrodynamic efficiency of the fish. The long *pectoral fins* are a unique feature of the albacore among the tuna family.

PELAGIC FISHES.

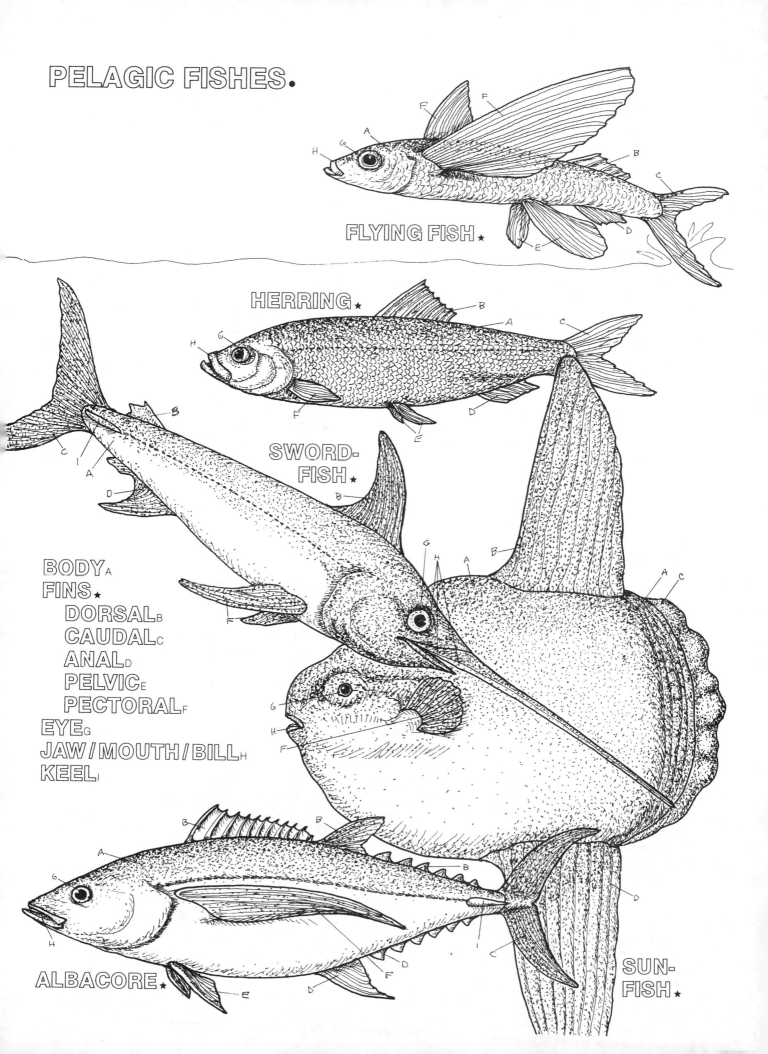

FLYING FISH ★

HERRING ★

SWORD-
FISH ★

BODY A
FINS ★
   DORSAL B
   CAUDAL C
   ANAL D
   PELVIC E
   PECTORAL F
EYE G
JAW / MOUTH / BILL H
KEEL I

ALBACORE ★

SUN-
FISH ★

Fish that live on, or very close to, the bottom (benthic fishes) do not need a streamlined body form for continuous swimming. This plate considers the body modifications of the tidepool sculpin, the sea robin, the stargazer, and the highly modified starry flounder.

Color each fish as it is discussed in the text. The body color varies in these species, but generally, they are a dark gray or brown in the stippled areas, and light colors elsewhere. Structures B through H are, again, the same colors as on the two previous plates. The electric organs of the stargazer are located under the skin; they should receive a separate color, and then be colored over.

The tidepool sculpin is a small (8 cm) fish, commonly found in Pacific coast tidepools of the United States. The sculpin forages for small crustaceans and other invertebrates among the rocks and algae of tidepools. Its coloration varies from mottled reds to greens and grays, and this mottled coloration breaks up its profile against the heterogeneous background. The sculpin usually sits motionless on its enlarged *pectoral fins,* but it can dart about rapidly in the water for short distances. Tidepool sculpins are representative of a number of small, elongate, relatively unspecialized benthic fishes (such as blennies, clinids, and gobies), all of which frequent rocky areas.

The sea robin has a large head, encased in an armor of rough, bony plates. Sea robins walk along the bottom using the first three rays of their *pectoral fins,* balancing the *body* using the remainder of the *pectoral* and *pelvic fins.* These three rays are articulated, and the fish uses them somewhat like fingers, pulling itself along and turning over rocks to uncover the crustaceans and molluscs on which it feeds. Sea robins are usually found on sand or sand–mud bottoms, at depths between 19 and 45 meters. Species occur in most oceans except the very coldest; they are gener-

ally small, although some large species grow to 1 meter in length. In places where they occur abundantly, sea robins are sometimes fished commercially.

The stargazer, a medium-sized fish (up to 51 centimeters), is a poor swimmer and usually lies buried in mud or sand. The Greeks called them "holy fish" because their small *eyes,* set on top of a rather square head, seemed to look toward the heavens. Some stargazer species have a fleshy lure attached to the floor of their large *mouth.* The lure attracts fish to within striking distance of the stargazers' *electric organs* whose discharge then stuns the prey. The *electric organs,* located behind the *eyes,* are thought to be derived from eye muscles and the optic nerve; they are capable of producing a stunning shock of 50 volts.

Another characteristic of these fishes is the presence of poisonous spines (not shown), located just behind the operculum and above the *pectoral fins.* These spines are grooved, with poison sacs at their base. Some species are believed to have a poison that can cause death in humans.

A far more benign, but equally specialized fish is the starry flounder which is a member of the family of left-eyed flatfish: both *eyes* are located on the left side of the fish. During its larval development, the right *eye* gradually migrates from its normal position to the other side of the *body,* while the *mouth* remains in the normal position. These fish are highly compressed and lie on one side on sandy bottoms, often with only their *eyes* and opercular opening uncovered. When the flounder swims, it does so in a sideways position. The starry flounder is very common in shallow, temperate waters of the Pacific, and is often found in nearly fresh water, especially when young. They grow to lengths of 90 centimeters and weigh as much as 9 kilograms (20 lbs). Starry flounders, unpigmented on their underside, get their name from the stellate scales that cover their body.

BOTTOM DWELLERS.

BODY A
DORSAL FIN B
CAUDAL FIN C
ANAL FIN D
PELVIC FIN E
PECTORAL FIN F
EYE G
JAWS/MOUTH H
ELECTRIC ORGANS I

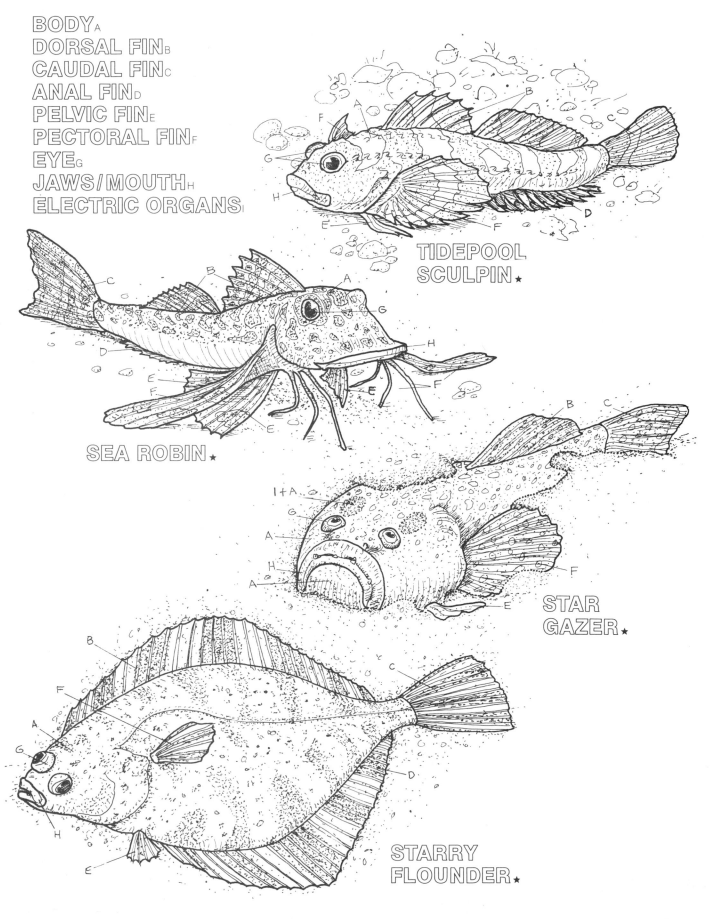

TIDEPOOL SCULPIN ★

SEA ROBIN ★

STAR GAZER ★

STARRY FLOUNDER ★

The pelagic zone and the homogeneous soft bottoms are monotonous environments compared to the coral reef, which is a tremendous mosaic of structure, form, and color. Few other marine habitats house such a variety of fish forms in such a small area as does the coral reef. The five fishes introduced here represent the diversity of reef fishes, but do not come near to exhausting the variety of fish forms living there.

The text recommends a specific color for each fish's body. Color each fish as it is discussed. Structures B through H are again colored the same as on the previous plates. As you color, study the variations of each structure, comparing them to the other fish on the plate.

The coral grouper is a pink or coral color. Groupers are commonly seen on coral reefs, swimming leisurely just above or among the coral structures. The grouper can quickly accelerate from its slow forward motion, and swim quite rapidly for short distances. Note its fusiform shape and broad, rounded *caudal fin*. Some species of grouper become quite large and fearless (230 kg or 500 lbs; 2.1 m in length). The grouper will readily approach divers and allow itself to be stroked and petted; they are easily trained to feed from a diver's hand.

The trumpetfish is an interesting fish sometimes seen swimming beside a grouper. This fish is long and slender (up to 76 cm), and yellow or orange in color. It is often found hanging motionless, head down, among the tall swaying coral growths or sponges. This posture is a deception; the trumpetfish is a crafty predator with the ability to dart, arrow-like, and seize a small fish in its *jaws*.

The golden boxfish stands out against the coral reef with its bright yellow or gold color. The boxfish displays what is known as warning coloration, advertising that its skin is covered by a poisonous secretion. Predators learn to associate this color with the taste, and thus leave the boxfish alone. Boxfish, also called trunkfish and cowfish, have a second line of defense: their entire *body* surface is underlain by bony plates that encase them in an unpalatable armor. This bone structure gives the fish a boxlike appearance, and renders it rather inflexible and a very poor swimmer. Lacking *pelvic fins,* boxfish swim mainly by the sculling action of the *anal, dorsal,* and *pectoral fins.*

The stiffness of the boxfish contrasts with the graceful agility of the threadfin butterflyfish. The color of this fish is gray or white. The butterflyfish is short (13–20 cm) and laterally compressed—a very maneuverable body shape. This fish can make extremely fast turns and quick stops; it can remain stationary in mid-water, back up, and move its head up and down. The *pectoral* and *pelvic fins* are important in these movements, and the fish uses its broad *dorsal* and *anal fins* to increase stability during fast turns. Butterflyfish swim around coral formations, poking in crevices and ledges for small invertebrate prey. When pursued, they dart in and out of these places with great facility, leaving faster, but less agile, predators behind.

Unlike the butterflyfish, the moray eel remains very close to its home. Eel *body* colors range from brown and green to gold. These fish have no *pelvic* or *pectoral fins,* and their *body* is long and snakelike. A *dorsal fin* runs the length of the *body,* and an *anal fin* extends from mid-body back, merging with the *dorsal fin* at the tail. Morays can swim in a slow, serpentine fashion, but are usually found entwined in the coral, under a ledge, or in a hole or cave. Their *mouths* are full of pointed teeth, but they will usually bite only when provoked or cornered. A few species are venomous. The prominent protruding *nostrils* are, as in most fishes, used only for smell. At night, morays leave their caves, crawling over the substratum, feeding chiefly on fish. Some of the larger species may grow to 2 or 3 meters in length.

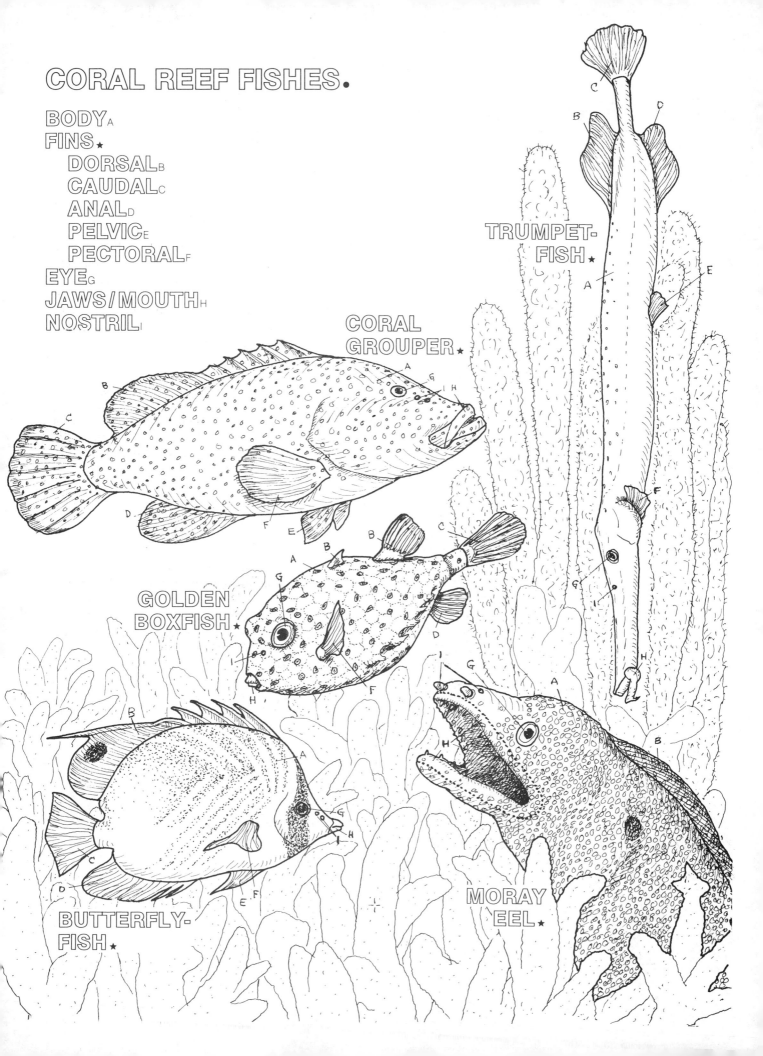

CORAL REEF FISHES.

BODY A
FINS ★
  DORSAL B
  CAUDAL C
  ANAL D
  PELVIC E
  PECTORAL F
EYE G
JAWS/MOUTH H
NOSTRIL I

TRUMPET-FISH ★

CORAL GROUPER ★

GOLDEN BOXFISH ★

BUTTERFLY-FISH ★

MORAY EEL ★

39

Sunlight penetrates the ocean to depths no greater than 300 meters, and usually much less, depending on the clarity of the ocean water. The greater volume of the ocean lies below this sunlit layer, and remains in constant darkness; the water temperature falls with increasing depth to a relatively constant 4 °C. Because plants cannot photosynthesize without light, there is little food in the depths of the ocean. Although such harsh conditions do not seem very favorable to life, a number of fishes are adapted to the mid-water and deep-sea habitats.

The middle layer of the ocean, where light fades into darkness, is known as the mesopelagic zone. Here, many fishes create their own light, using it in intraspecific communication, to capture prey, and for other purposes. The deepest ocean habitat is called the bathypelagic zone. The bathypelagic fishes are mostly very small and have enormous mouths with long, sharp teeth that enable them to capture and swallow large meals. This plate presents six mid-water/deep-sea fishes as an introduction to these fascinating mesopelagic and bathypelagic creatures.

Color each fish as it is mentioned in the text. The three mesopelagic fishes in the top half of the plate should receive a light gray color; the three bathypelagic fishes should receive a dark gray. The light-emitting organs can be colored blue-green, the color of light produced by them.

Because the upper layers of the water column hold an abundant food supply, many species undertake a nightly migration to the surface to feed. The lanternfishes are one such group of migratory species. Lanternfishes are so named for the row of bioluminescent *light organs* (photophores) along their ventral surface. These *light organs* are used in intraspecific communiction, allowing lanternfish to recognize potential mates, for example. Lanternfish have very large *eyes,* presumably an adaptation to night feeding in dimly lit waters.

The Pacific viperfish feeds on lanternfish and squid, and follows them upward in their nightly migration. Also equipped with *light organs,* the viperfish has an oversized *mouth* and fanglike teeth. In order to feed, the viperfish must tilt its head upward and extend its lower *jaw* to a point where its *gills* are exposed. The viperfish is small (22-30 cm) and has an elongated *dorsal lure* that may be used to entice prey.

The hatchetfish is a mesopelagic fish that remains in waters several hundred meters below the surface. The upward-oriented *eyes* of the hatchetfish scan the faintly lit waters for zooplankton. The hatchetfish has binocular vision (somewhat unusual in fish) which allows it to gauge the position of prey hovering above. The light organs are used in an ingenious defensive behavior known as counter-lighting (Plate 56).

The black devil, also called anglerfish, has a large *light organ* mounted on a modified dorsal fin spine, which serves as a lighted *lure* and is held above the *mouth.* Prey is attracted to the light and quickly engulfed by the black devil's tooth-filled *mouth.*

Another deep-sea fish with an extremely large *mouth* is the pelican gulper eel, which can reach a length of 76 centimeters. The gulper eel has a distensible gut, and will often consume prey (primarily crustaceans) as large as itself.

The tripodfish lives in the deep ocean from the North Atlantic to the Caribbean Sea. It is considered a benthic fish, feeding on small zooplankton that hover just above the bottom. The "tripod" consists of the elongated *pelvic* fins and the ventral lobe of the *caudal* fin, and allows this fish to feed in the water column without having to expend energy in swimming. Zooplankton bump into the threadlike extensions of the tripodfish's fins, alerting the fish to their presence. The tripodfish grows to about 25 centimeters and has tiny *eyes.*

MID-WATER AND DEEP-SEA FISHES.

BODY A
FINS ★
    DORSAL B
    LURE B¹
    CAUDAL C
    ANAL D
    PELVIC E
    PECTORAL F

EYE G
JAW/MOUTH H
LIGHT ORGAN I

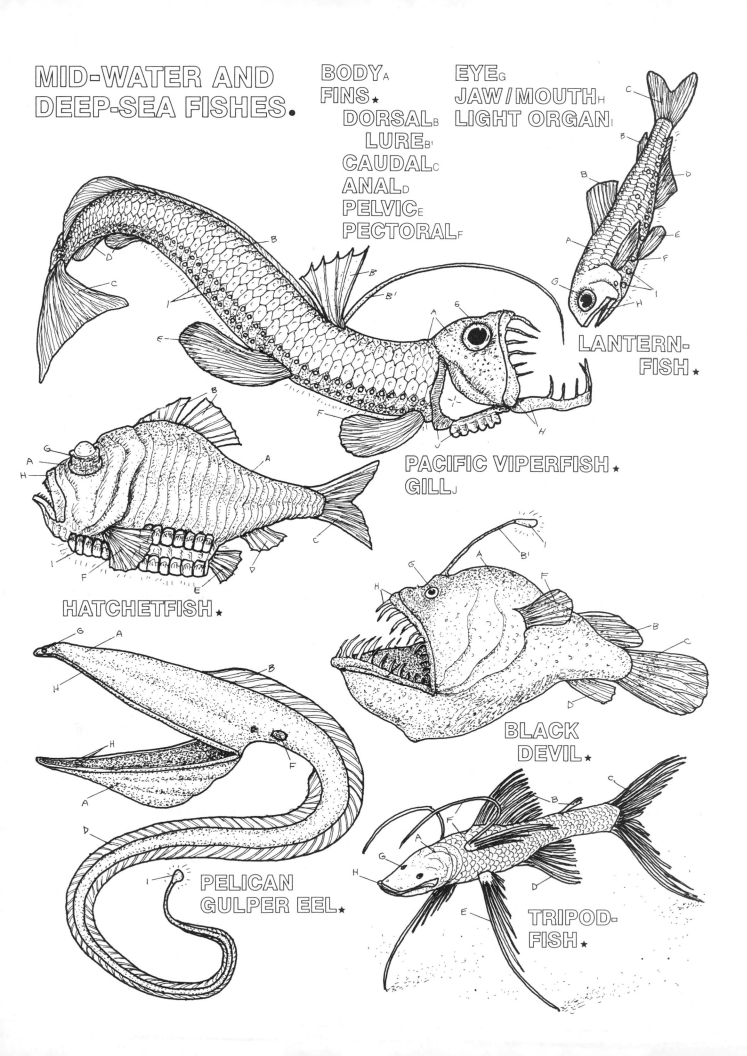

LANTERN-FISH ★

PACIFIC VIPERFISH ★
GILL J

HATCHETFISH ★

PELICAN GULPER EEL ★

BLACK DEVIL ★

TRIPOD-FISH ★

40

BONY FISH AND SHARK: COMPARISON OF STRUCTURE

The previous five plates dealt with the variety of forms found in the bony fishes. Before venturing forward in a study of the cartilaginous sharks and rays, a comparison of these two groups is in order. The major organs and systems, with special emphasis on the difference between the sharks and rays and the bony fishes, are reviewed here.

This plate presents a highly diagrammatic comparison between the internal anatomy of a bony fish and that of a shark. Locate and color each organ or system in both illustrations as it is discussed in the text. Only those structures labeled and outlined with dark lines are to be colored.

Starting from the anterior, or mouth end of each fish, note the relatively small *brain,* which continues posteriorly as the *spinal cord.* The large *olfactory lobe* of the brain gives evidence of the importance of an acute sense of smell in both groups of fishes. The *olfactory lobe* terminates near the base of a blind sac, which opens at the *nostril.* Since the *nostrils* do not open to the throat — as they do in mammals for instance — most fishes must take in their respiratory water current through the mouth. Rays and other bottom-dwelling elasmobranch fishes take in water through the *spiracle.* Contraction of the throat musculature pumps water over the *tongue* and across the *gills,* which take up the sides of the throat. This oxygen-laden water passes over *gill filaments,* oxygenating the blood supply. *Gill rakers* serve to spread the *filaments* apart for maximum surface exposure, and to clear foreign matter from the filaments. The water is pumped out of the gill chamber past the operculum in bony fishes, or through the gill slits in sharks.

The relatively small *heart* is located near the base of the *gills.* It pumps blood through the *gills* and from there to the head and the rest of the body. *Kidneys* help regulate blood chemistry and deliver waste products to the exterior through the urogenital opening (not shown). The *gonads* (sex organs) also empty through this opening.

Another large organ linked to the circulatory system is the *liver,* whose main functions are to store surplus nutrients and to detoxify certain substances. In sharks and their relatives, the *liver* has an additional function: to contribute buoyancy to the body. This is because the *liver* stores oil which is considerably less dense than water. The presence of the oil in the shark *liver* is responsible for the latter being much larger than the *liver* of the bony fish. Most bony fish utilize a more efficient adaptation for buoyancy: the gas-filled *swim bladder.* Air is either gulped at the surface or secreted from the bloodstream into the *swim bladder.* Delicate regulation of the gas content in this organ allows a fish to maintain its position in the water column with minimum expenditure of energy. By contrast, elasmobranchs function similarly to airplanes, requiring forward motion to keep from sinking.

The length and complexity of the gut has more to do with the diet of any particular species than with that species' relation to either major group. The *spiral valve* is one gut structure found almost exclusively in sharks and their relatives. It serves to increase the surface area of the *intestine* for more efficient absorption of nutrients.

The characteristic difference between the two major groups, by which they are most commonly named, is the skeletal material. Elasmobranchs are cartilaginous fishes. Their skeletons are made of a relatively flexible material, but the skeletal structure is much less elaborately articulated than in bony fishes. The result is that the bodies of elasmobranchs are on the whole less maneuverable and adaptable than those of bony fishes. This is especially apparent in the structure of the fins and the uses to which they are put.

Most fishes are covered by a protective layer of scales. In elasmobranchs, these are *placoid scales,* also called denticles. The word "denticle" indicates a relation to teeth, and sharks' teeth indeed originate from this same layer of tissue. Bony fishes possess, as a group, several types of scales. The *ctenoid* scales shown here overlap to provide both protection and suppleness.

COMPARISON OF STRUCTURE:

BRAIN_A
OLFACTORY LOBE_A1
SPINAL CORD_A2
NOSTRIL_B
TONGUE_C
GILL_D
GILL FILAMENT_D1
GILL RAKER_D2
HEART_E
KIDNEY_F

GONADS_G
LIVER_H
STOMACH_I
INTESTINE_J
ANUS_K
MUSCLE_L

BONY FISH.

SWIM BLADDER_M
BONE_N
VERTEBRA_N1
RIB_N2
FIN SUPPORT_N3
CTENOID SCALE_O

SHARK.

SPIRACLE_P
SPIRAL VALVE_Q
CARTILAGE_R()
VERTEBRA_R1
PLACOID SCALE_S

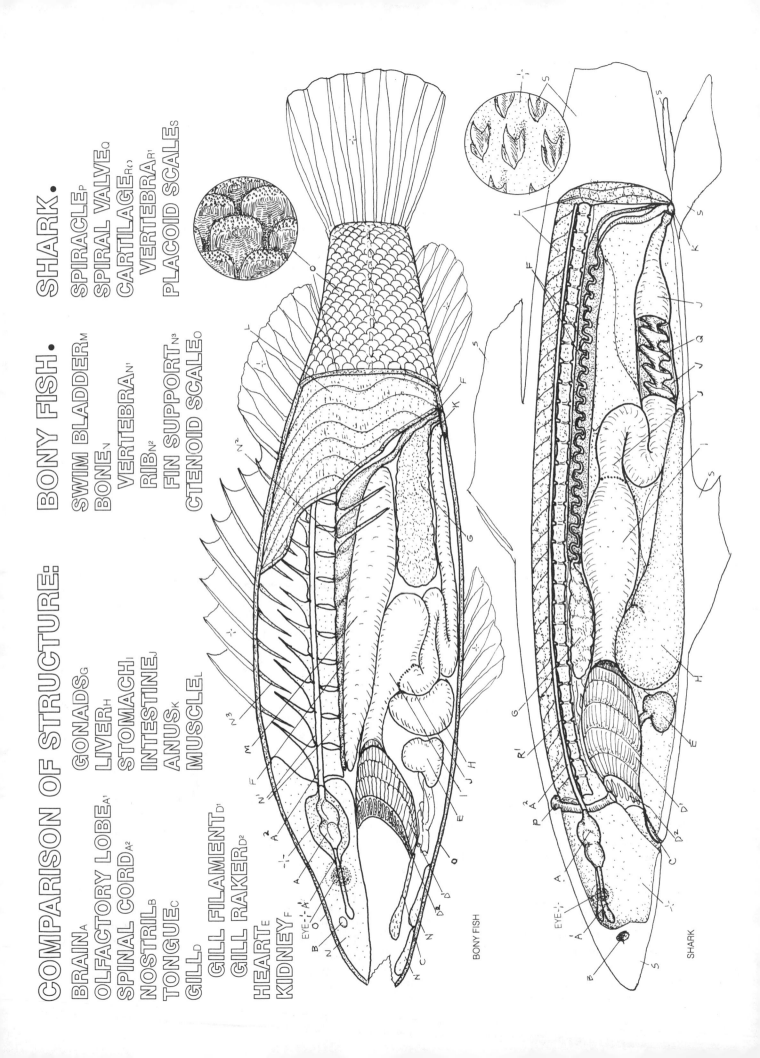

BONY FISH

SHARK

The elasmobranch fishes (sharks, skates, and rays) are characterized by flexible, cartilaginous skeletons. They differ from bony fishes in two other major ways: They possess gill slits instead of an operculum, and they lack a swim bladder. These characteristics have a profound influence on the mode of existence of the elasmobranch groups. This plate introduces the three elasmobranch forms.

Color each animal separately as it is discussed in the text. When coloring the dogfish's head, notice that the area of the mouth/jaws is bounded by a dotted line. The upper right inset drawing illustrates a single row of teeth in the lower jaw moving forward with use and being discarded.

The spiny dogfish shark (1 meter) possesses a streamlined *body* with a complement of both paired *(pectoral* and *pelvic)* fins and unpaired *(dorsal, anal,* and *caudal)* fins similar to the bony fishes (see Plate 35). However, there are some differences. The *dorsal fin* of sharks is more rigid and is incapable of being folded down flush against the back. While the spiny dogfish possesses two *dorsal fins,* they do not straddle the midline and are not properly considered as "paired." The *caudal fin* is usually asymmetrical (heterocercal), with a larger upper lobe. This larger lobe, supported nearly to its tip by the vertebral column, gives the shark an upward, as well as forward movement that counteracts its tendency to sink. Similarly, the large *pectoral fins* are more rigid than those of bony fishes and are held horizontally with a slight upward cant that provides a hydroplanelike lift. The *pelvic fins* of the male each have an elongated *clasper* on the inside; this is used in fertilization of the female (see Plate 70).

The absence of an operculum means that sharks cannot actively pump water over their gills, and this suggests that they have to swim constantly to maintain a flow of oxygen-bearing water over their gill surfaces. However, many bottom-dwelling sharks possess an opening *(spiracle)* from the dorsal body surface to the gills through which the latter can be ventilated. The *spiracle* is fitted with a non-return valve that opens and closes as the fish breathes; water is drawn in through it and expelled out the *gill slits.*

The shark's *mouth* is usually underslung, or subterminal, with a pointed snout projecting above it. When a shark strikes, it raises its snout and projects its wide-open *jaws* forward, allowing it to take a substantial bite. Shark *teeth* are usually pointed and sharp. As they break off or are worn down, they are continuously replaced from behind, as the arrows in the inset indicate. Unlike other vertebrates, the *teeth* of sharks are mounted in skin, not jaw bone. Shark *teeth* are actually modified placoid scales, or *denticles,* the same basic structure that covers the shark's body, giving it a texture like sandpaper. *Teeth* are lined up in rows of five or six (fewer in some cases; more in others) and grow forward as the leading *tooth* wears or falls out.

Color the underside (ventral side) of the big skate.

The skate is a much flattened (depressed) elasmobranch that spends its daylight hours buried in the sand on the bottom, with only the eyes and large gill-ventilating spiracles uncovered. At night the skate emerges and swims, using undulations of its long winglike *pectoral fins.* The skate feeds on bottom-dwelling crustaceans and molluscs that it captures in its ventrally located *mouth* and crushes with its flat, blocklike teeth. The skate's *body* is moderately slender and often has rows of enlarged *denticles* along the back. The male's *pelvic fins* possess long *claspers* for mating. The *anal* and *caudal fins* are absent, and the *dorsal fins* are very small. Skates are found in all seas, and species range in size from one meter to the "barndoor" skate which is over three meters long.

Color the stingray, including the enlarged spine at lower right.

The southern stingray inhabits Atlantic and Caribbean waters. Stingrays are similar to skates in the shape of the *pectoral fins,* position of the *mouth, spiracle,* and *gill slits,* and in behavior and diet; they spend the daylight hours buried in soft substrata and the nighttime foraging for bottom-dwelling crustaceans and molluscs. Rays differ from skates in having a long whiplike tail that lacks dorsal fins and possesses one or more *spines* at its base. The *spines* are modified *denticles* and, like the shark's *teeth,* are replaced in series when broken or pulled out. Buried in the sand, these rays are almost invisible, and the unwary wader who steps on one is often impaled on one of the *spines* by the lashing motion of the tail. Stingray "stings" are complicated by the presence of venom glands on each side of the *spine;* the venom flows along grooves in the *spine* and into the wound. The pain is often excruciating, and some stings have proven fatal to humans.

INTRODUCTION TO FORMS.

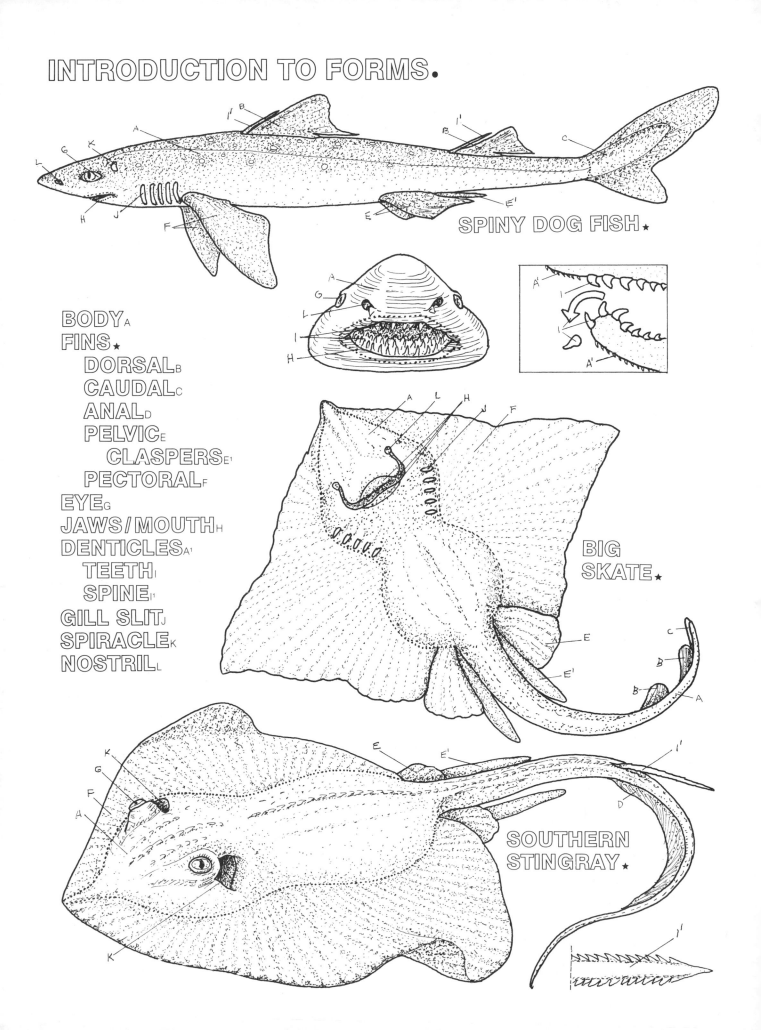

SPINY DOG FISH ⋆

BODY A
FINS ⋆
    DORSAL B
    CAUDAL C
    ANAL D
    PELVIC E
       CLASPERS E'
    PECTORAL F
EYE G
JAWS / MOUTH H
DENTICLES A'
    TEETH I
    SPINE I'
GILL SLIT J
SPIRACLE K
NOSTRIL L

BIG
SKATE ⋆

SOUTHERN
STINGRAY ⋆

There are about 250 species of sharks, and many of these are widely distributed in the world's oceans. This plate introduces some of the unique or more notorious members of the group.

Begin with the basking shark and color each animal separately as it is dicussed in the text. The arrows indicate the feeding current in the basking shark. Note that the lower half of the great white shark is not to be colored in either drawing so as to give the natural appearance of countershading. Note jaws/mouth of the hammerhead do not show in the larger view, and it and the thresher shark lack a keel. Only the upper drawing of the basking shark is labeled. Match the labeled parts on the basking shark with the corresponding parts on the other sharks.

The basking shark is the second largest fish in the sea, reaching lengths of 9 meters or longer. The filter-feeding basking shark is depicted with its cavernous *mouth* open, its typical feeding position as it swims slowly along, engulfing large zooplankters and small fish. Water passes out through the large *gill slits* while small prey are caught on the gill raker surfaces, then swallowed. Basking sharks are thought to be harmless to humans. Travelling in schools of up to 100 individuals, basking sharks are found in the temperate zones of the Atlantic and Pacific on both sides of the equator. They are sometimes hunted for their oil-rich liver because of its high vitamin content.

The large (6 meters) hammerhead shark has all the typical shark characteristics except for the head, whose flattened cephalic lobes give it a rectangular shape suggesting a hammer. The *eyes* and *nostrils* are located at the tip of these lobes. As the hammerhead swims, it swings its head back and forth through the water. This behavior is thought to increase the likelihood that the shark's sensory receptors will

detect food, by broadening its swath through the water. In addition, the flattened head may act to increase the shark's maneuverability. Hammerheads are found in tropical and subtropical waters of all oceans, and some attacks on humans have been recorded.

The great white shark has a cosmopolitan, warm water distribution, ranging into temperate waters such as are found off California. The great white accounts for the majority of verified shark attacks on humans, but the reasons for this are not agreed upon. Some experts believe great whites are territorial and see humans as intruders in their territory. Others feel divers are mistaken for seals or other marine mammals; and still others feel that the movements of swimmers, mistaken for those of distressed fish, attract the shark. Perhaps the most plausible theory is that the great white shark is mainly curious and, unlike most sharks that "feel" an unknown substance by bumping it with their snouts, the great white "feels" with its *mouth*. Even a gentle "feel" or taste with its tremendous, tooth-studded *jaws* can be fatal to a human swimmer. The great white is recognizable by its large black *eyes* and its lunate *caudal fin,* which is not the typical shark heterocercal form. The great white is also stouter than most sharks. A record great white, over 6 meters in length, was recently taken off Cuba. Such a shark would weigh well over 1,600 kilograms (3,500 lbs).

The thresher shark is immediately recognized by the tremendously elongated dorsal lobe of its *caudal fin*. The *caudal fin* may account for up to one-half the shark's body length, which may reach 6 meters; threshers weigh up to 450 kilograms. Threshers are found in the warm and subtropical waters of the Atlantic and Pacific oceans, often feeding in groups on schools of small fish. They swim circles around their prey, using the large *caudal fins* to stun and scare the fish into a group that can be fed upon more easily ("threshing behavior").

SHARKS.

BODY A ★
FINS ★
    DORSAL B
    CAUDAL C
    ANAL D
    PELVIC E
    PECTORAL F
EYE G
JAWS/MOUTH H
GILL SLIT I
KEEL J
NOSTRIL K

BASKING SHARK ★

HAMMERHEAD SHARK ★

GREAT WHITE SHARK ★

CEPHALIC LOBE

THRESHER SHARK ★

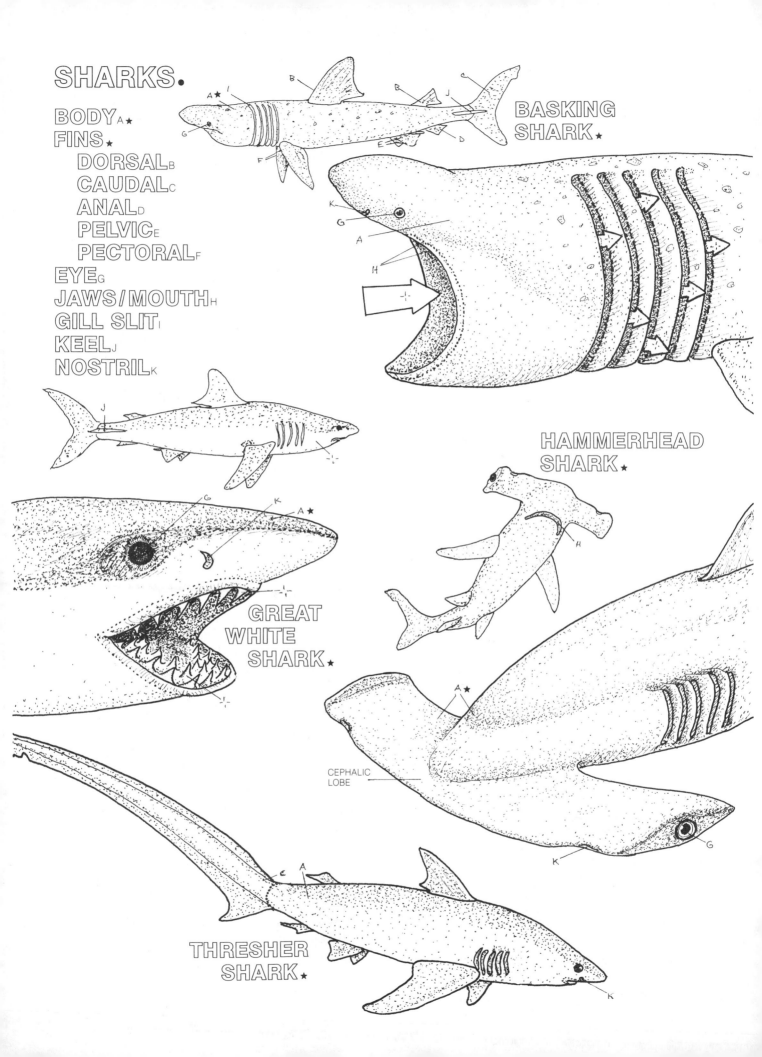

43
ELASMOBRANCH FISH DIVERSITY: RAYS AND RELATIVES

Color the manta ray at the top of the page.

The Atlantic manta ray, like all mantas, possesses special *cephalic fins* on either side of its *mouth*. These fins are the anterior portion of the *pectoral fins* and serve to form a funnel through which plankton and small fish are directed into the *mouth* of the filter-feeding manta. Because the *cephalic fins* are widely separated and held in front of the *body,* they were called "horns" by fishermen who referred to the manta as a devilfish. Mantas are graceful creatures which swim by slowly beating their huge, winglike *pectoral fins.* The Atlantic manta shown here may reach a width of over 6 meters and weigh in excess of 1,360 kilograms (3,000 lbs). Mantas are renowned for their habit of leaping high into the air and coming down like a shot; this may serve to loosen parasites or perhaps to stun fish.

Now color the spotted eagle ray.

The spotted eagle ray is found inshore in all tropical and warm temperate seas. This large, active ray (over 2 meters in length; 230 kilograms) is generally seen swimming near the bottom joined in formation with several other rays. Its long tail (more than three times its body length) is characterized by from one to five spines near its base and is held straight out behind the *body.* The placement of the *spines* makes them a questionable defensive weapon compared to the posteriorly located spines of other stingrays. The eagle ray's *body* is thicker than other stingrays, and the head projects forward from the disc formed by the *pectoral fins* and *body.* The head is thick and boxlike and harbors powerful jaws with pavementlike teeth used in crushing bottom-dwelling crustaceans and molluscan prey. The *eyes* and *spiracles* are mounted on the sides of the head rather than on top.

Large numbers of eagle rays often enter estuaries at high tide to feed on commercial shellfish. They use their *pectoral fins* like plungers ("plumber's friend") to pop shellfish out of their burrows; their passage can leave a mudflat looking like a minefield. Oystermen often construct fences and other deterrents to foil the marauding rays.

Next color the sawfish.

The sawfish looks distinctly un-raylike. However, it shares with the rays a bottom-dwelling existence and ventrally placed gill slits. Its form is dominated by the long, paddle-like extension of the head that bears pairs of *teeth* (actually modified denticles like shark teeth and stingray spines). The sawfish uses its "saw" to dig shellfish out of the bottom and to stun fish. It will swim through a school of fish slashing wildly and impaling or stunning fish, which it then eats. Sawfish also use their weapon defensively against natural predators and humans.

The *body* of the sawfish is decidedly more sharklike than raylike, and it swims by using its *caudal fin* in typical shark fashion. These animals can be quite large (over 6 meters in length and 365 kilograms) and are found inshore in all tropical seas. They often enter brackish and even fresh water, as demonstrated by a large population of sawfish landlocked in Lake Nicaragua.

Now color the electric ray. The electric organ is located beneath the skin and is given a separate color for illustrative purposes.

The giant electric ray is thick and flabby compared to other rays, and its almost circular *body* is smooth and without scales. The *body* is continuous with a relatively thick tail characterized by two *dorsal fins* and a distinct *caudal fin.*

Two large kidney-shaped *electric organs* are located on the *pectoral fins,* just above the *pelvic fins* on both sides of the *body.* These *organs* are modified muscle tissue directly innervated by the brain. They contain special hexagonal tissues that are, in essence, storage batteries wired in series. The *electric organs* account for one sixth of the ray's total weight and can generate an initial shock in excess of 200 volts, enough to electrocute a large fish.

Ichthyologists diving at night off southern California reefs have found that the electric ray emerges from its half-buried, daylight repose and actively hunts along the reefs. When it physically encounters other fish it shocks them with electricity and then uses its *pectoral fins* to move the stunned, inactive fish into its ventrally located mouth.

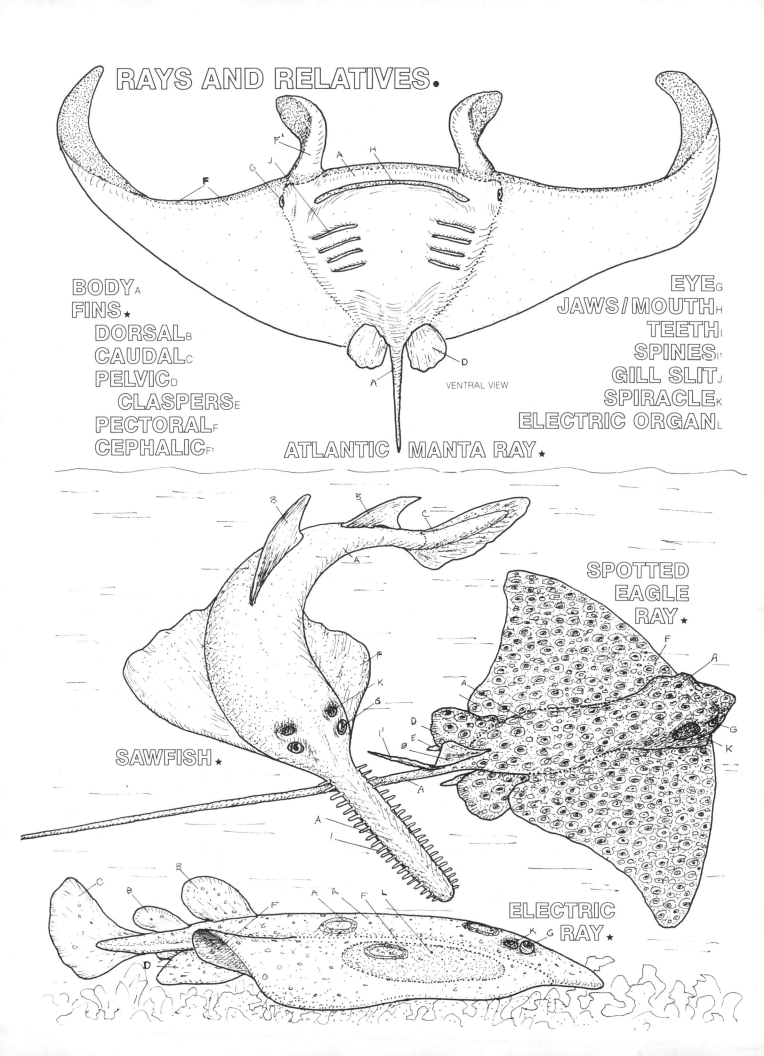

RAYS AND RELATIVES.

BODY A
FINS ★
DORSAL B
CAUDAL C
PELVIC D
CLASPERS E
PECTORAL F
CEPHALIC F¹

EYE G
JAWS / MOUTH H
TEETH I
SPINES I¹
GILL SLIT J
SPIRACLE K
ELECTRIC ORGAN L

VENTRAL VIEW

ATLANTIC MANTA RAY ★

SPOTTED EAGLE RAY ★

SAWFISH ★

ELECTRIC RAY ★

44
MARINE REPTILES: TURTLES AND SEA SNAKES

In terms of both numbers of individuals and numbers of species, fishes are by far the most successful group of vertebrates in the sea. Reptiles, on the other hand, are represented by only a few marine species. Most of these are still tied to the land by the necessity of having to lay their eggs there. In this plate, two reptiles are introduced: the green sea turtle and the *yellow-bellied sea snake.*

Begin by coloring the adult green sea turtles at the top of the page. Color the four illustrations of egg laying, hatching and tracks. Note that the broad tracks receive the same color as the forelimbs that created them.

The green sea turtle, an endangered species of the Caribbean, is one of several species of turtle that spend their lives in the sea. We know more about the green turtle than the others because it is of some economic importance. Many products are made from its shell, and the meat is considered a delicacy.

Studies by Dr. Archie Carr of the University of Florida have focused on the green turtle rookery (breeding beach) of Tortuguero, in northeastern Costa Rica, where a large number of turtles come to reproduce. Once every three years an adult female green turtle undertakes a journey back to the beach where she was hatched to lay her own *eggs.* For some turtles, these migrations may be several hundred miles long. Males and females mate in the surf just offshore from the rookery. The male grasps the female with his large *forelimbs* and transfers his sperm to her as shown in the upper drawing. After a few days, the female makes a nocturnal trip onto the beach. She pulls herself up the beach with her *forelimbs,* all the way to the dry sand of the upper beach. She digs a broad pit with her *forelimbs* and then delicately excavates a bottle-shaped *burrow* with her agile *hind limbs* (center illustration, far left). The female lays approximately 100 leathery-skinned *eggs* in the *burrow* and carefully covers them with sand. She buries the pit entirely and throws sand all about to disguise the location of the nest. Her job completed, the female returns to the sea. Her broad *tracks* left behind indicate the difficulty this turtle has in moving on land (center illustration,

second from left). The *forelimbs* are modified into highly effective swimming flippers, but they cannot lift her bulk off the sand. Similarly, her *carapace* is much reduced and streamlined for swimming; it does not serve as a fortresslike retreat, unlike those of many freshwater and terrestrial turtles.

Before leaving the breeding grounds the female may return to the beach as many as five times at 15-day intervals. The *eggs* incubate in the warm sand for about 60 days, and the young hatch all at once and begin to dig to the surface (center illustration, second from right). They emerge at night and instinctively find their way to the ocean (center illustration, far right). The green turtles are most vulnerable to predation during their time in the *burrow* and during their scramble to the sea.

The young turtles remain at sea and do not reappear in the sea grass beds until at least one year later. When 4 to 6 years old, the females will return to the exact stretch of beach where they hatched and contribute to the next generation.

Color the title "yellow-bellied sea snake" a golden yellow. Color the light portion of the snake the same color, and the dark pattern black. Note that the flattened tail receives a different color.

Sea snakes are found in shallow tropical and subtropical waters. All are related to the cobra family. They have a potent venom that can cause severe injury to humans. The *yellow-bellied sea snake* is found in the Pacific off the coast from Panama to Mexico. It is commonly seen on the surface, often in aggregations of several hundred individuals.

Sea snakes are well adapted to a marine existence. Many give birth to the young alive at sea, and the newborn snakes can immediately swim on their own. The sea snake has a *flattened tail* used as a paddle in swimming. Sea snakes typically feed on fish, and can remain submerged for thirty minutes or longer between breaths. Most species are docile, although some attacks on divers have been reported. They are best appreciated from a distance.

TURTLES AND
SEA SNAKES.

GREEN SEA TURTLE ★

HEAD A
FORELIMB B
TRACK B'
HIND LIMB C
CARAPACE D
TAIL E
EGG F
BURROW G
JUVENILE H

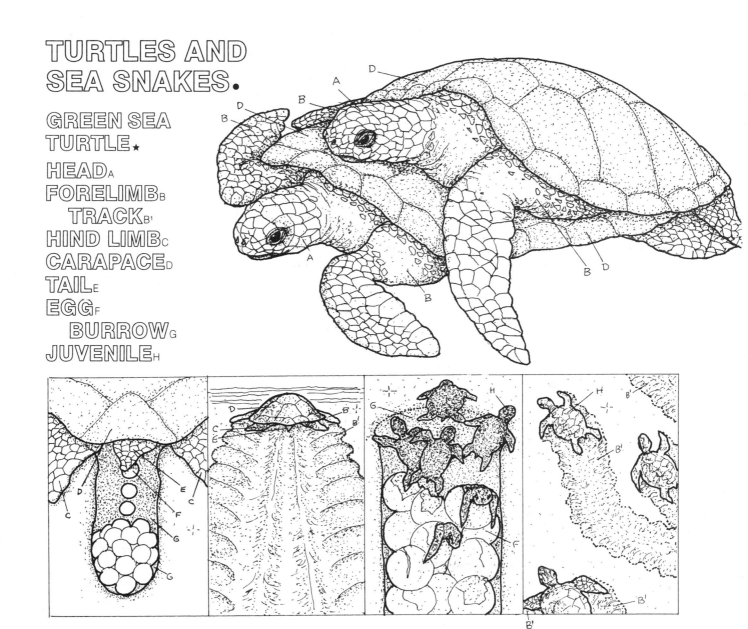

YELLOW-BELLIED SEA SNAKE I
FLATTENED TAIL J

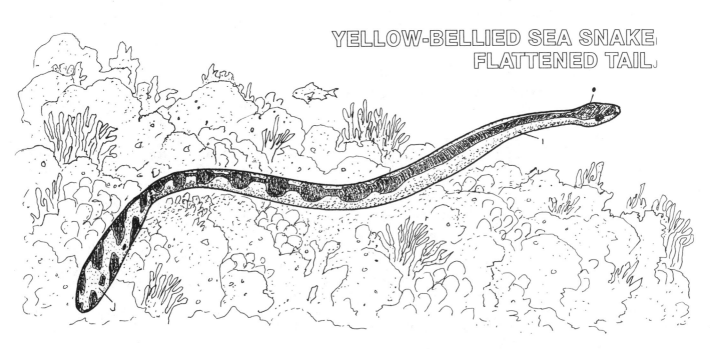

MARINE MAMMALS: INTRODUCTION TO FORMS

Four or more groups of warm-blooded mammals have invaded the seas during evolutionary history. In a relatively short time (fifty million years), one of these groups has evolved into the largest creatures in the sea: the great whales. In this plate, a representative of each of these four groups is discussed.

Color the sea otter first. Color the body as indicated in the text to simulate natural coloration.

The least specialized marine mammal is the sea otter; it is also the least changed from its ancestral terrestrial form. Sea otters are closely related to the smaller river otters, and both are classified in the weasel family, along with badgers, wolverines, and minks.

Sea otters have retained all four limbs although they are somewhat modified for a life spent almost entirely in the water. The *forelimbs* are short and have stubby, rounded paws with poorly developed fingers; however, sea otters are able to pick up rocks from the bottom and use them as tools. The *hind limbs* of the otter are short, and the feet are large and webbed. Otters swim using their webbed feet and *tail,* which is flattened dorsoventrally to provide a broader sculling surface. Sea otters usually swim on their back. They also rest, sleep, and eat on their back, with all four limbs projecting out of the water, possibly to conserve body heat. Unlike most other marine mammals, sea otters do not possess a layer of insulating blubber and must rely on the air trapped in their fine, dense fur to maintain warmth. This fur, which varies in color from reddish brown to black, is highly prized by furriers, and fur hunters nearly rendered the sea otter extinct. Sea otters are relatively large. Male otters often grow to 1.35 meters in length, including tail (25–35 cm), and weigh over 36 kilograms.

Now color the sea lion. As with all mammals shown here, color the outline illustration of the entire animal as well.

The California sea lion represents the pinniped order, which includes the "earless" seals, sea lions, and the walrus. Their closest terrestrial relatives are the bear and dog. The sea lion's *forelimbs* are modified into *flippers,* which it uses to "fly" through the water. The *hind limbs* are also flipperlike, though much less powerful, and the sea lion possesses a short stub of a *tail.* The sea lion's tawny, sleek form is evidence of the streamlining that evolved for ease in swimming. A

layer of blubber beneath the skin provides insulation against the cold ocean water.

Now color the dugong.

The dugong of Africa and the South Pacific, and its close Atlantic relative, the manatee, are representative of the sea cows. These animals, which grow to 3 meters and can weigh over 400 kilograms, have little hair and rely on a layer of blubber for insulation. Supposedly, sailors mistook them for mermaids, and thus their scientific name (Sirenia) derives from the comely Sirens who lured Odysseus' crew onto the rocks in Greek mythology.

The dugong, whose closest terrestrial relative is the elephant, is the only herbivorous mammal restricted to the sea. It spends its entire life in the water, where its chief food is sea grass. It uses its bristled muzzle to root up tender young sea grass plants from shallow protected bays and estuaries. In Florida, manatees perform a service to navigation by eating the troublesome water hyacinth that clogs slow-flowing river channels.

The *forelimbs* of the dugong are flipper-like and the hind limbs are absent. The *tail* has grown into broad, notched flukes, which the dugong moves in powerful beats, propelling itself along at maximum speeds of thirteen miles per hour. Although they may appear sluggish, dugongs are actually quite alert and active, and are said to have an intelligence comparable to that of a deer. Although the dugong lacks ear flaps and has only a small opening to the outside, it is able to hear quite well.

Finally, color the dolphin.

The familiar bottlenose dolphin is a representative of the order Cetacean which includes the whales and dolphins. These animals exhibit the most thoroughgoing adaptation to marine existence of all the sea mammals. Like the dugong, the Cetaceans lack external ear flaps, and have lost their hind limbs, relying on horizontally positioned *tail flukes* for propulsion. The dolphins also possess a rigid, permanently erect *dorsal fin* for swimming stability. The sleek *body* of the dolphin is extremely efficient hydrodynamically; the animal is known to attain swimming speeds of twenty-four miles per hour. Like all marine mammals, both dolphins and whales must breathe air, and their *nostrils* are equipped with muscular plugs that close when they dive.

INTRODUCTION TO FORMS.

BODY A
FORELIMB B
　　FORE FLIPPER B¹
HIND LIMB C
TAIL D
　　FLUKE D¹
EYE E
EAR F
NOSE/NOSTRIL G
DORSAL FIN H

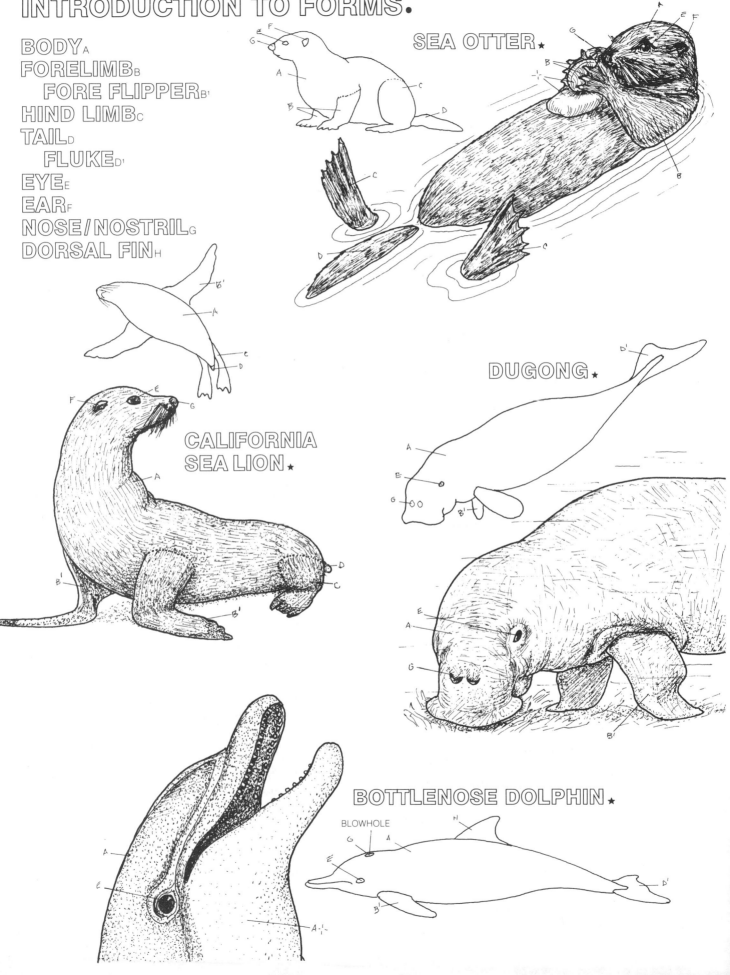

SEA OTTER ★

CALIFORNIA
SEA LION ★

DUGONG ★

BOTTLENOSE DOLPHIN ★

BLOWHOLE

The pinnipeds ("feather-foot") are the most visible of the marine mammals because of their large size and their frequenting coastal areas where they rest and sun themselves. In this plate, the fur seal, harbor seal, walrus, and elephant seal are introduced.

Color each animal separately as it is discussed. Use a light color for the body so as not to conceal the detail of the drawings.

In the illustration, a male and female northern fur seal are shown "hauled out" on a rocky platform. These animals are representative of the "eared" seals (including the sea lions) so called because they possess an obvious *ear* flap, or pinna. Fur seals can rotate their *hind limbs* forward to rest underneath them and support their *bodies*. Their *forelimbs* are relatively large and strong enough to allow them to hold their upper *bodies* erect; they can move fairly well on land.

Northern fur seals have very fine, rich fur that keeps them dry and warm. The large male is dark brown, and the smaller female is dark gray. The northern fur seal is found from Baja California to the Bering Sea, and rarely comes on land, except to breed. Each year 50,000 seals are killed on the Pribilof Islands in the Bering Sea.

The upright mobility of the fur seal on land contrasts strongly with the harbor seal's awkwardness out of water. Harbor seals belong to the family of true seals, which also includes the elephant seals. These seals lack the external ear pinna and have fairly restricted movement on land. Their *hind limbs* do not protrude from the *body* trunk above the ankle; therefore they cannot rotate under the *body* when on land but instead drag behind. The *forelimbs* are relatively small and do not support the upper *body* well. As a result, a seal's movement on land consists of flopping along on its belly. In the water, the *hind limbs* propel the harbor seal along. When held close to each other, they may act in tandem in a sculling motion.

Harbor seals are found in both the Atlantic and Pacific oceans and appear to be less nomadic than other pinnipeds. They stay close to shore and often haul out on sandbars in bays and estuaries. Harbor seals feed mainly on small fish, molluscs, and crustaceans in near-shore waters.

The elephant seal is the largest of the pinnipeds. Males may be over 6 meters long. Females are much smaller, reaching 3.5 meters. The most prominent feature of the male elephant seal is its enlarged *nose*, or *proboscis*, which is a secondary sex characteristic. As the male becomes sexually mature, the *nose* begins to grow, achieving full development at, it is thought, eight to ten years of age. During mating season, the *nose* is used as a display organ while adult males (in the posture shown) bellow challenges at each other. For many of the pinnipeds, mating season involves very complex social interactions (see Plate 75).

The walrus is found in the cold, northern waters of both the Atlantic and Pacific, near the edge of the Arctic ice pack. The male walrus may reach a length of 3.6 meters, and the female is only slightly smaller. Both sexes possess prominent ivory *tusks*. These enlarged teeth were once thought to serve as digging devices for rooting out shellfish from the bottom. However, more recent evidence suggests they use their broad muzzle for this chore, with the stiff whiskers serving as tactile sensory devices that detect prey. It now appears that *tusk* length may be related to social status.

At birth the walrus has a thin coat of reddish hair; the skin of adults is nearly smooth. The walrus relies on a blubber layer for warmth; it spends much of its time hauled out on ice floes where it sleeps and rests. These large animals have only one major predator besides humans—the polar bear, which eats mainly young walrus. The Pacific walrus population has enjoyed a remarkable recovery from its slaughter by hunters in the nineteenth and early twentieth century, but it is still very vulnerable to human predation.

PINNIPEDS.

BODY A
FORELIMB B
HIND LIMB C
TAIL D
EYE E
EAR F
NOSE G
     PROBOSCIS G'
TUSK H

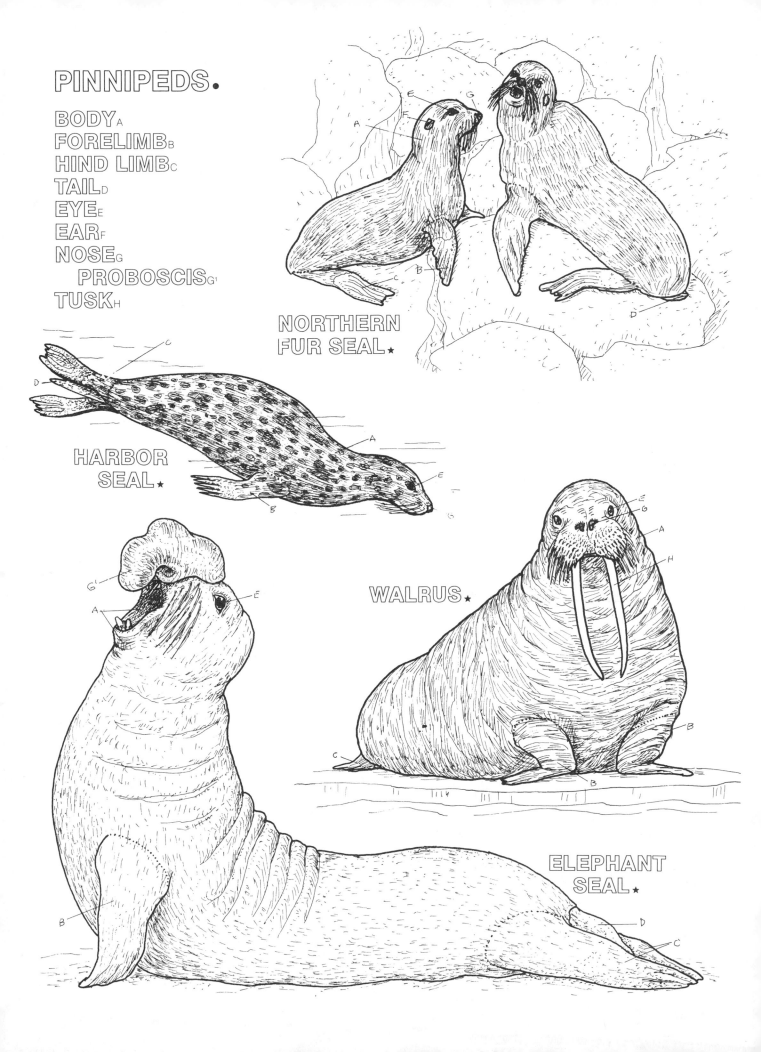

NORTHERN
FUR SEAL ★

HARBOR
SEAL ★

WALRUS ★

ELEPHANT
SEAL ★

47

MARINE MAMMALS: TOOTHED WHALES AND ECHOLOCATION

There are approximately seventy-four species of toothed whales, including porpoises, dolphins, sperm whales, beaked whales, and a number of other cetaceans (Order Cetacea), all of which bear teeth and have a single blowhole. Dolphins differ from porpoises in having longer beaks, although the names are often used interchangeably. The toothed whales are all predators and eat fish, squid, and, in one or two cases, other marine mammals. They have lost their sense of smell and have fairly good vision, although sight is of little use to the deeper-diving forms that probe the murky depths for prey. The best-developed sense in the toothed whales is hearing, which is used by some species in a very sophisticated behavior known as echolocation.

Begin by coloring the porpoise and the dolphin. Note that the jaw of the dolphin receives a separate color. The melon is an internal structure exposed for identification.

Echolocation has been most thoroughly studied in the smaller toothed whales, especially the dolphins. Underwater, these animals emit a tremendous range of vocalizations from squeals, chirps, and moans to pulses, or trains, of very short "clicks." The former vocalizations are perhaps used in communication, while the latter "clicks" are used in echolocation. The dolphin is believed to emit from the *blowhole* a series of split-second clicks that may be repeated as often as 800 times per second. Some scientists believe these clicks are focused into an *outgoing* directional *beam* of sound pulses by the *melon,* a large, lens-shaped organ made of fatty tissue, located on the dolphin's forehead. When the clicks strike a *target,* a portion of the *signal* is *reflected* back. The bony *lower jaw* receives and transmits the impulse via bone conduction to the bone-enclosed inner ear. Here the impulses are converted to nerve impulses, and these are directed to the brain. The dolphin is able to determine the distance to a target on a continuing, moment-to-moment basis apparently by "measuring" the time between the emission of the clicks and their return. The rate of click production is regulated to allow the returning echo to be heard between outgoing clicks. Using this amazing skill, the dolphin can determine the size, shape, direction of movement, and distance of an object in the water. It scans the water column with low-frequency clicks and uses higher-frequency clicks to make finer discriminations.

Color the sperm whale. The spermaceti organ

is an internal structure exposed for identification. Also color the squid prey.

The sperm whale is the largest of the toothed whales, with males exceeding 17 meters in length and 47,000 kilograms (52 tons). Sperm whales got their name from whalers who hunted them for their spermaceti oil, used at first in lamps and later for lubricating precision machinery. A large amount of this oil is contained in the huge *spermaceti organ* located in the outsized snout (weighing up to 11,000 kilograms or 12 tons) of the sperm whale. This organ is physically very complex, consisting of a mass of oil-filled connective tissue surrounded by layers of muscle and blubber. There is speculation that this huge organ may be concerned with buoyancy, or it may be analogous to the *melon,* focusing the vocalizations possibly used in echolocation.

As in the dolphin, the origin and the actual mode of reception of the echoed clicks in the sperm whale is still not completely understood. The sperm whale produces very loud low-frequency clicks (believed to be emitted from the *blowhole*) that travel for miles. Some believe these loud clicks allow the sperm whale to scan the depths for its *squid prey.* Once the prey is located, the sperm whale dives for it. Dives have been recorded that reached depths of 1,134 meters (3,700 feet) and lasted 90 minutes.

Finally, color the killer whale and its prey. In the lower right corner the toothed whales are drawn in relative scale with the blue whale, whose length can reach 33 meters.

The killer whale is the largest of the dolphins (9 meters in length), and, as in the sperm whale, the male is considerably larger than the female. Killer whales are favorites of oceanariums because of their intelligence and ease of training. They show true affection for their human playmates, and their great leaps thrill audiences. Some caution is advised, however, for the killer whale sports a mouthful of large conical teeth that are closely spaced and interlock when the mouth is closed. Killer whales are skillful hunters and their *prey* includes other marine mammals, such as seals, sea lions, sea otters, porpoises, and even the great baleen whales. They hunt in cooperative schools and have been observed literally stripping the flesh from the biggest of all animals, the blue whale. Killer whales are found worldwide from the tropics to the polar regions.

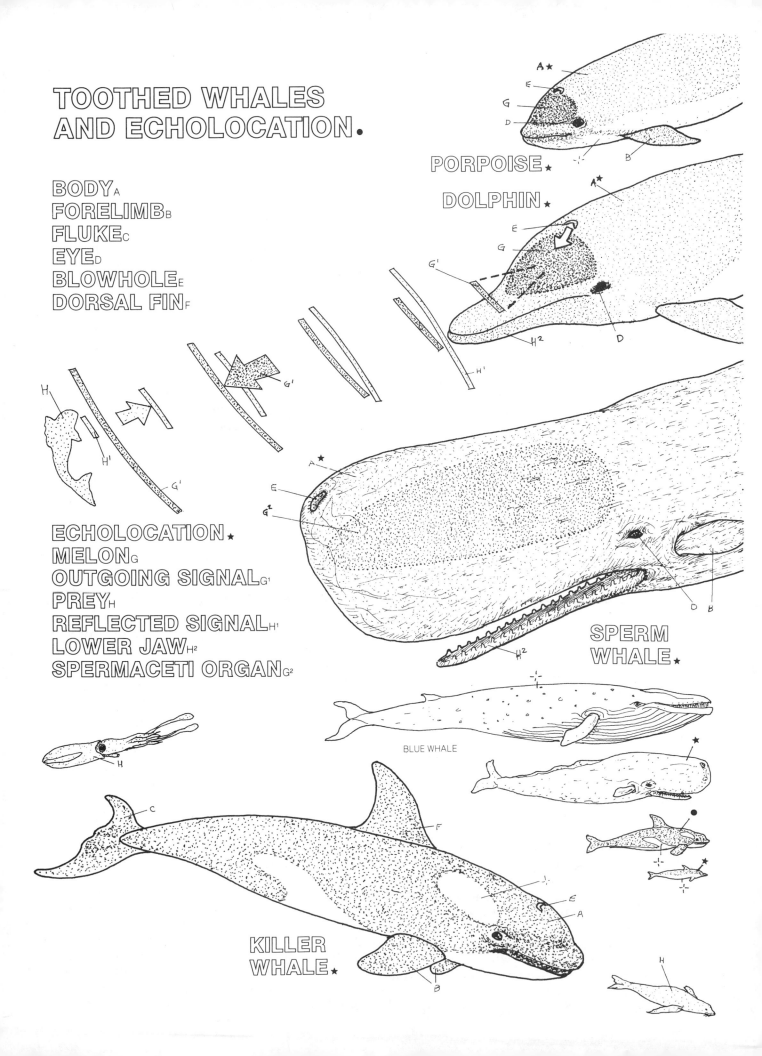

TOOTHED WHALES AND ECHOLOCATION.

BODY_A
FORELIMB_B
FLUKE_C
EYE_D
BLOWHOLE_E
DORSAL FIN_F

ECHOLOCATION ★
MELON_G
OUTGOING SIGNAL_{G1}
PREY_H
REFLECTED SIGNAL_{H1}
LOWER JAW_{H2}
SPERMACETI ORGAN_{G2}

PORPOISE ★

DOLPHIN ★

SPERM WHALE ★

BLUE WHALE

KILLER WHALE ★

48

MARINE MAMMALS: BALEEN WHALES

The baleen whales are the largest animals that have ever lived on earth. Yet little is known about them. On account of present-day whaling activities, some species may well disappear before we ever have a chance to learn more about them. Baleen whales differ from the toothed whales in several ways including shape of the head, presence of baleen plates, lack of teeth and absence of the large melon in the forehead.

Begin with the upper drawing. Note the bones of the forelimb. Note the single detached baleen plate drawn forward of the upper jaw.

Baleen whales are named for the fibrous, horny *baleen plates* hanging from their *upper jaws.* These *plates* are entirely different from teeth and are derived, like hair, from the epidermal tissues. The individual *baleen plate* is about 6 millimeters thick and consists of long, coarse fibers held together by a horny substance. These *plates* are lined up along the *upper jaw* at intervals of about 7 millimeters. The outer surface of each *plate* is smooth, while the fibrous bristles of the inner surface are frayed and cover the spaces between the *plates.* These frayed bristles make an effective filter for straining sea water. The shape and size of the *baleen plates* and the coarseness of the bristles vary among whale species.

Color the right whale. Use a dark gray for the whale's body, as this whale is nearly black in color.

The right whale (so designated by early whalers as the right whale to kill because it did not sink) is the name used for two closely related species that have the most elaborate baleen apparatus. It has a significant arch in the *upper jaw* from which hang *baleen plates* up to 4 meters in length. The *lower jaw* is not similarly arched, but the large lower lips extend well above the *lower jaw* to enclose the *baleen.* The flexible *baleen* plates fold backward toward the throat when the mouth is closed and spring forward when the mouth opens. The head of the right whale takes up nearly one-third the length of the animal's *body.*

Right whales feed on copepods and other planktonic crustaceans that accumulate in large patches or shoals on and beneath the ocean's surface. To feed, the whale simply swims through the plankton patch with its scooplike mouth open, forcing the water through the *baleen plates* and trapping the plankton on the bristles. The black right whale is found in all oceans between the Arctic and Antarctic circles, and reaches a length of 20 meters.

Color the humpback whale.

The humpback whale and its fellow great whales, or rorquals (fin, sei, blue, minke, and Bryde's whales), are distinguished from other baleen whales by the presence of many long, pleated grooves in the throat region. Humpbacks are characterized by their long, knobby *forelimbs,* and are known for the elaborate songs that are sung each year by the males on the breeding grounds.

The *baleen* of the humpback and other rorquals is short and broadly based in the roof of the mouth. These whales feed on a variety of planktonic crustaceans; some species also take fish and squid. The whale swims through the plankton patch, and its capacious mouth engulfs several tons of seawater. Its huge tongue is used as a piston to force water out over the *baleen* where the small crustaceans are trapped and drawn into the throat to be swallowed. Some recent observations of feeding behavior suggest the humpback swims upward in a slow spiral while emitting air from its *blowhole.* This forms a circular curtain of bubbles which apparently startles the planktonic crustaceans. In this manner, they are driven to the circle's calm center, where they are then engulfed by the humpback.

Color the gray whale. The animal is illustrated swimming on its side while feeding just above the bottom. The patchy areas on the body are barnacle encrustations and should be also colored gray.

The gray whale is the only member of its family. Gray whales are medium-sized (15 meters), active whales found in the North Pacific (see Plate 74). Their *baleen* is the shortest of all the whales, and their mode of feeding is quite unique. Gray whales appear to feed primarily on benthic crustaceans, especially amphipods. The method seems to be that they swim just off the bottom, turn on their sides, and sweep their heads back and forth; this disturbs the amphipods and causes them to rise off the bottom. Then the gray whale sucks the amphipods in under the *baleen* by pushing its tongue against the floor of its mouth while expanding the throat grooves to create suction. This behavior has been seen only in captive animals. However, the gray whale's *baleen* is usually shorter on one side of its mouth, and that same side is also relatively free of the barnacles that encrust the hides of these gentle giants. Both observations suggest that the behavior just described (and illustrated at the bottom of the plate) is a typical mode of feeding for the gray whale.

BALEEN WHALES.

BODY A★
FORELIMB B
FLUKE C
EYE D
BLOWHOLE E
DORSAL FIN F

HEAD ★
    UPPER JAW G
    BALEEN PLATE H
    LOWER JAW I
    TONGUE J

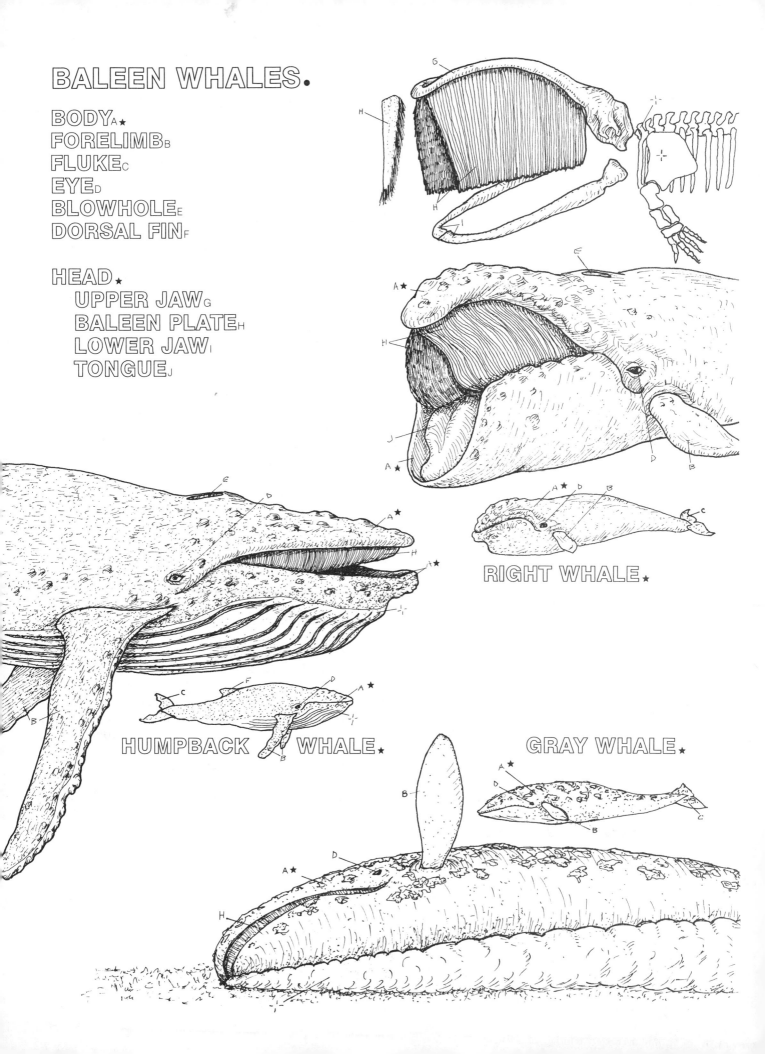

RIGHT WHALE.★

HUMPBACK WHALE.★

GRAY WHALE.★

49

COLORATION IN FISH: THE ADVERTISERS

Color in fishes (and in many other animals) serves a variety of important functions. These will be investigated in the following seven plates, which offer an opportunity to employ some of your brightest colors. This plate is concerned with fishes that call attention to themselves: that is, they advertise.

Color the garibaldi and lionfish. Color the entire garibaldi, including fins, a golden orange or golden yellow, also color the center portion of the traffic light on the fish's right. Color the dotted, striped areas of the lionfish red. Don't be too concerned about accuracy in coloring the confusing pattern. Color the skull and crossbones red.

One type of advertisement is seen in the bright, golden orange *garibaldi* of southern California. This fish uses solid, bright color to advertise its presence, warning other *garibaldis* not to encroach on its closely guarded territory. Many brightly colored reef fishes are thought to use their colors in this kind of territorial display.

Warning coloration also functions between species (interspecifically), as in the case of the golden boxfish in Plate 38. The butterfly *lionfish* of the tropical Pacific has showy red and white stripes on its body, pectoral fins, and highly venomous dorsal spines. The contrasting red and white colors give warning to predators, especially those that have felt the sting of those poisonous spines before. Interspecific advertisement is not always negative; one example being that of the cleaner wrasses (Plate 77), which advertise their availability with bright color patterns.

Now color the angelfish and the butterflyfish. The larger angelfish on the left receives a yellow color in areas marked C, as does the officer's insignia in the drawing on the right. The areas labeled D are given a bright blue color. For the juvenile angelfish, use bright blue D in the striped areas that are drawn in light. The heavily lined stripes are left blank. The remainder of the smaller fish is colored black. In the copperband butterflyfish the stripes marked E receive a pink- or orange-tinted copper color. Note that the well-advertised (but false) eye is colored black. The false bottom of the bottle represents false advertising in that the apparent amount in the full bottle is not, in fact, the true amount.

Coloration may also play a role in species recognition, especially when many closely related fish species are living near one another, such as on a Pacific coral reef. Many fish have a different color pattern in their juvenile phase, sometimes strikingly so. Illustrated here is the *Koran angelfish,* which as a juvenile is black with blue and white semicircles, and as an adult is light yellow with darker yellow spots and light blue on the fins and operculum. These distinct color patterns allow quick recognition and serve to clarify social behavior within a species.

Coloration can make a fish appear to be something it isn't. Eye-bars, broad stripes, and other tricks are often employed in such "false advertising." The beautiful *copperband butterflyfish* possesses a black eyespot (ocelus) near its tail which potential predators mistake for an eye. The predator will "lead" its prey (just as a human hunter will lead a moving target) and plan its strike for the head, only to gobble empty water or a small piece of dorsal fin that the *butterflyfish* can regenerate. The eyespot trick is employed by several types of fish and appears throughout the animal kingdom in groups as diverse as moths and octopuses.

THE ADVERTISERS.

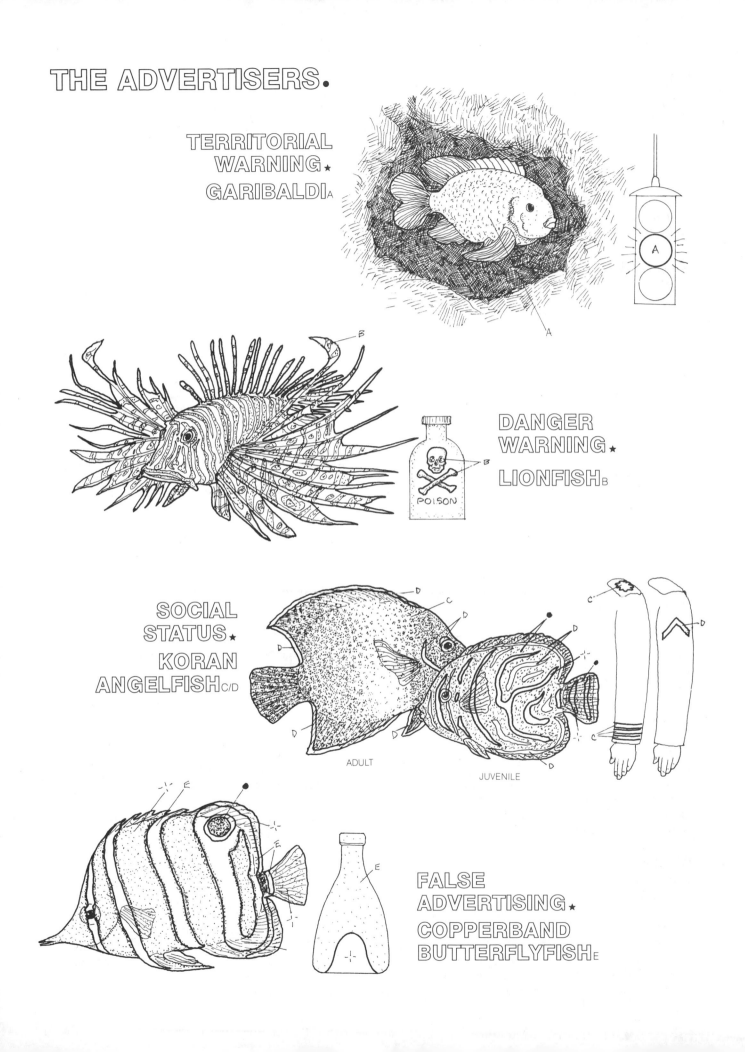

TERRITORIAL
WARNING ★
GARIBALDI A

DANGER
WARNING ★
LIONFISH B

SOCIAL
STATUS ★
KORAN
ANGELFISH C/D

ADULT

JUVENILE

FALSE
ADVERTISING ★
COPPERBAND
BUTTERFLYFISH E

POISON

50

COLORATION IN FISH: THE CRYPTIC ONES

Color the countershading of the mackerel in the drawing at top right (above the rectangle) and the related title dark blue (A). Do not color its white or silvery underside. On the left color the grouper and related title orange (B). Now color the rectangle representing sea water blue-green (D), but do not color the fish. Then color the top half of the grouper (above the dotted line) a light orange (B), or leave small areas within the colored area white. This is to suggest the lightening of the top half of the fish by light shining down from the surface. Color the bottom half solid orange (B) and then color over the bottom half with gray (C★) to simulate the darkening of the bottom side that would be in shadow. Color the mackerel the same blue-green as the water on top (D), and shadow the bottom surface (C★).

There are several modes of cryptic coloration among fish. Many fish of the upper pelagic zone, such as the common mackerel of the North Atlantic, seen here, have a coloration pattern known as obliterative countershading. These fish are green to blue-black on their dorsal surface and grade into a silver to almost white underneath. Viewed from above, they blend with the dark void below; seen from below they blend with the bright, sunlit surface water. Viewed from the side (upper illustration), it is apparent that the surface normally directed toward the light is *countershaded* by the dark color on the fish's back, while the ventral (belly) surface, which would normally be in shadow, is counterlighted. The sides have tones that grade between these colors. As a result, the fish is rendered optically flat and reduced in visibility. This type of countershading is employed not only by fish, but also by many birds, mammals, reptiles, and amphibians. Compare the mackerel's color pattern with that of the *non-countershaded* grouper (upper left). One can easily see how the grouper, which inhabits coral reefs, would stand out in open water.

Within the block labeled "disruptive coloration," color the anemonefish (including fins), to the right, bright orange (B¹), but do not color the two stripes on the body of the fish or the tubular coral surrounding it. On the grouper, to the left, color the dotted patches and stripes (E) brown and the body (F) tan or dull yellow. Color the coral marked G red, H pink, E brown, and I dull green. Color the sea water blue-green (D).

Another type of color pattern widely employed by fish is disruptive coloration. A familiar object is recognized by a specific contour or outline that shows an obvious surface continuity. The orange garibaldi against a dark bottom is a good example. Disruptive coloration conceals a particular recognizable form by employing contrasting colors in different-sized patches that effectively break up and distract from a recognizable outline, such as the camouflage employed on military equipment. Fish that live in a habitat with a high degree of surface relief, like a coral reef or a kelp forest, take advantage of its many shapes, shadows, and colors to blend into the background.

The grouper on the left has patches of color in irregular shapes and positions all over its body, including its fins and lips. Standing alone against a solid background it would be easily recognized. However, when the fish remains still against a highly colorful and irregular coral reef habitat, the shadows and light areas tend to blend with the fish's coloration, and it no longer presents a clear outline to the viewer. Furthermore, groupers are capable of quickly changing their color patterns and background colors to match the changes in lighting and habitat encountered as they swim.

The clown anemonefish also employs disruptive coloration. When seen against a solid background, the small fish's body, vivid orange with boldly contrasting white stripes, is clearly obvious. However, among the tubular coral and finger sponges of coral reefs, the broad white stripes blend with the background, effectively disrupting the outline of the fish and making it much less recognizable.

Color the stonefish (E) brown in the middle of the bottom drawing and the surrounding rocks (E¹) brown with various red accents (G) that are surrounded by dark lines. Color the bottom (F) tan or dull yellow.

The deadly stonefish mimics and blends into its rocky background with the aid of its coloration. Its body is lumpy and unstreamlined to further conceal its motionless presence among the encrusted stones, from which it cannot be distinguished by humans or prey fish (aggressive mimicry). Like its relatives the lionfish and scorpionfish, this fish has extremely venomous spines; they have caused death to humans who inadvertently stepped on them while wading on shallow Pacific coral reefs.

THE CRYPTIC ONES.

OBLITERATIVE COUNTERSHADING ★

NON-COUNTERSHADED: COUNTERSHADED A

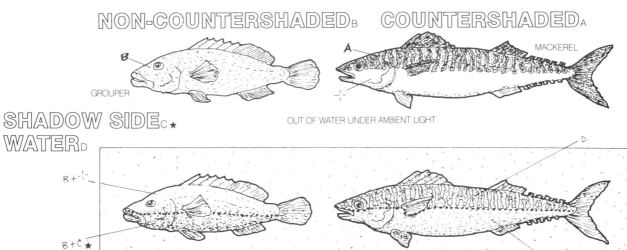

B GROUPER

A MACKEREL

OUT OF WATER UNDER AMBIENT LIGHT

SHADOW SIDE C ★
WATER D

B + ¡ -

B + C ★

D

D

D

C ★

AS SEEN IN THE WATER WITH LIGHT FALLING FROM ABOVE

DISRUPTIVE COLORATION ★

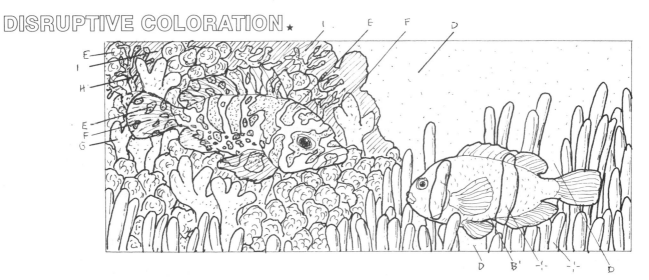

E
I
H
E
F
G

I
E
F
D

D
B'
- ¡ -
- ¡ -
D

AGGRESSIVE MIMICRY ★

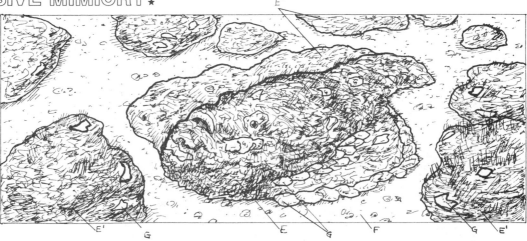

E'

E'
G
E
G
F
G
E'

This plate explores the source of color in fishes and how coloration can be changed in certain situations.

Begin with the patches of skin at the bottom of the page and color the base color of the skin (A) either a pale yellow, tan or light gray. Then color each of the color cells as follows: (B) red, (C) orange, (D) yellow, and (E) black. When coloring the title: "Color Cells (chromatophores) B, C, D, E," and "Pebbly Bottom B, C, D, E," you can randomly color the various letters with any of these colors. Then color each of the expanded chromatophore cells in the same manner (B, C, D, E). Note that both squares below are of the same magnification and represent an enlarged view of skin chromatophores.

In the large illustration of the flatfish, begin with the left half, coloring the skin of the fish (A). This is usually very close in color to that of the sandy bottom, which receives the same color. This base color is also the base color of the fish in the pebbly environment, and, since there is sand between the pebbles, color over that area with the base color (A) first. From here on, use B, C, D, or E in any combination you wish on both the fish and the pebbly structure. To get the optimum effect, if you do not want to color the entire illustration, color small adjacent sections of each illustration completely. Note that you should not use the chromatophore colors on the fish on the sandy bottom (to the left) since the pigmented areas are so highly concentrated that they cannot be seen.

Fishes take their color from two types of pigment cells that are located in various layers of the skin. One type of cell, the iridocyte (mirror cell) contains guanin, a substance that reflects light and color from the environment outside the fish. The iridocytes give rise to the pearly white color and the silvery, iridescent blues and greens often seen in fishes.

The *chromatophore* is a second type of color cell, which contains its own pigment particles of red, orange, yellow, and black. The cell body itself is highly branched, and in order for color to be seen the pigment granules must be dispersed throughout the branches. When the pigment is concentrated in the center of the cell, very little color shows. Both conditions are illustrated at the bottom of the plate. Fishes can produce other than the pigment colors by activating a mixture of *chromatophore* types. Green, for instance, can be obtained by combining black and yellow *chromatophores*.

Fishes change color depending on the color of their surroundings, the stage of their life cycle, or even a state of excitement. A startled fish will often blanch, the colors appearing to wash out of its skin as the pigment concentrates in the *chromatophores*. An angry fish may turn bright red as the pigment disperses in its red *chromatophores*. Some fishes change color pattern between night and daytime. A courting male fish may put on a "suit" of dazzling nuptial colors to attract a mate.

The best-studied color change in fish deals with their response to surroundings. These may change due to variations in incoming and reflected light or because the fish moves from one habitat to another. Many fish are able to assume the color of their background, and some are actually able to mimic its pattern. Flatfish are especially adept at this. As shown in the drawing, the flatfish concentrates the pigment in its *chromatophores,* enabling it to blend with the sandy bottom. When it moves to a pebbly bottom, clusters of *chromatophores* expand their pigments to match the size of the pebbles encountered.

Some fish change color with startling quickness, while others take minutes or even days. The color change is controlled either through the nervous or the endocrine (hormonal) system. The rapid changes are controlled by direct neural impulses to the *chromatophores,* while slower changes are brought about by blood-borne pituitary hormones. Short-term changes in color involve the concentration and dispersion of pigment granules in established *chromatophores*. Longer-term changes, such as those brought about by a permanent or long-term change in surroundings, are the result of an increase or decrease in the number of *chromatophores*.

The ability to register and recognize mood, sexual readiness, or social status by color has allowed fish to develop highly complex behavior patterns that are still only partly understood.

CHANGING COLOR.

BASE COLOR OF SKIN A
COLOR CELLS (CHROMATOPHORES) B, C, D, E ●

SANDY BOTTOM A¹   PEBBLY BOTTOM B, C, D, E ●

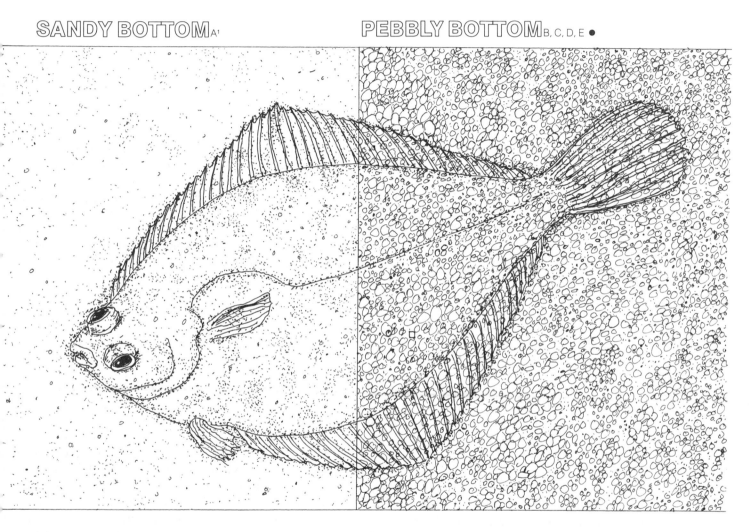

CONTRACTED PIGMENT ★   EXPANDED PIGMENT ★

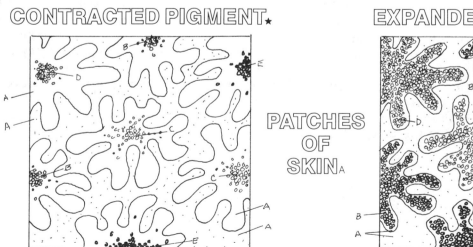

PATCHES
OF
SKIN A

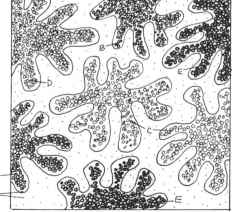

52
COLORATION IN MARINE INVERTEBRATES: THE ADVERTISERS

As we have seen, many fishes display various colors to attract attention to themselves. The same is true of some marine invertebrates.

In this plate you should use precisely the colors indicated so that the animals shown will look as lifelike as possible. Begin by coloring the peppermint shrimp. The antennae and the stripe down the middle of the back are white and should be left blank. The areas (A) adjacent to the white stripe are bright red, and the underside (B) of the shrimp is golden yellow. You may wish to color the title "peppermint shrimp" alternating red and white (blank).

The *peppermint shrimp* is a cleaner of fishes, and its brightly patterned body apparently advertises this service. This Caribbean shrimp resides in cracks and crevices of coral reefs, where its long white antennae and red-and-white striped body are visible to passing fishes. Once attracted, a fish presents itself to the cleaner, often using complicated patterns of behavior, and the shrimp emerges to pick parasites from the fish's body. Some observers believe that this and other cleaning shrimps (see Plate 77) actually establish routine "cleaning stations" which may be frequented daily by their customers.

The female cuttlefish (the left member of the courting pair) in the middle of the plate displays a brown or tan (C) background with dark brown (D) spots. The male on the right shows dark brown (D) stripes on a white (blank) background. Apparently to confuse prey, an attacking cuttlefish can rapidly change colors from dark brown (D), to white (-¦-), to a mottled pattern of light brown (C) splotches with dark brown (D) centers. The nudibranch at the bottom of the page should be colored blue-purple (E) with yellow-orange (F) stripes. Color the titles with the alternating colors indicated.

The *cuttlefish* is a master of color change. Like its relatives, the squid and octopus, the *cuttlefish* can change color in a fraction of a second (see Plate 54) for a variety of purposes, two of which are described here. *Cuttlefish* tend to be rather solitary except during the mating season, when the males begin to search out females. Identifying females could be a real problem were it not for courtship behavior involving color changes. The male, sporting a distinctive zebra-striped pattern over its body, approaches another *cuttlefish*; if the prospective mate is another male it responds by presenting a similar pattern. A brief skirmish may ensue until one swims off. However, when the approached *cuttlefish* is a female and does not assume the striped pattern, the male initiates mating by grasping its mate head-on as shown in the illustration.

The *cuttlefish* also uses color changes to confuse or dazzle potential prey. As the *cuttlefish* moves toward its intended victim it rapidly changes color patterns from dark to light to mottled. Thus distracted and taken off guard, the prey animal is captured by the *cuttlefish's* two long tentacles.

The warning colorations displayed by various *nudibranchs* (sea slugs) have been discussed earlier (Plate 25). The example shown here is brilliant purple with contrasting yellow stripes, seemingly advertising itself as a potential meal. It turns out that many *nudibranchs* possess some noxious attribute that a predator might well learn to associate with their distinctive color patterns. Some species (like the one illustrated here) secrete distasteful or irritating chemicals on their bodies; some have stinging nematocysts imbedded in their surface; and still others may accumulate noxious substances from their prey. However, in some species no such explanations for their bright colors have been found; perhaps they are mimicking some distasteful look-alike.

THE ADVERTISERS.

PEPPERMINT SHRIMP A+B
CUTTLEFISH C+D
NUDIBRANCH E+F

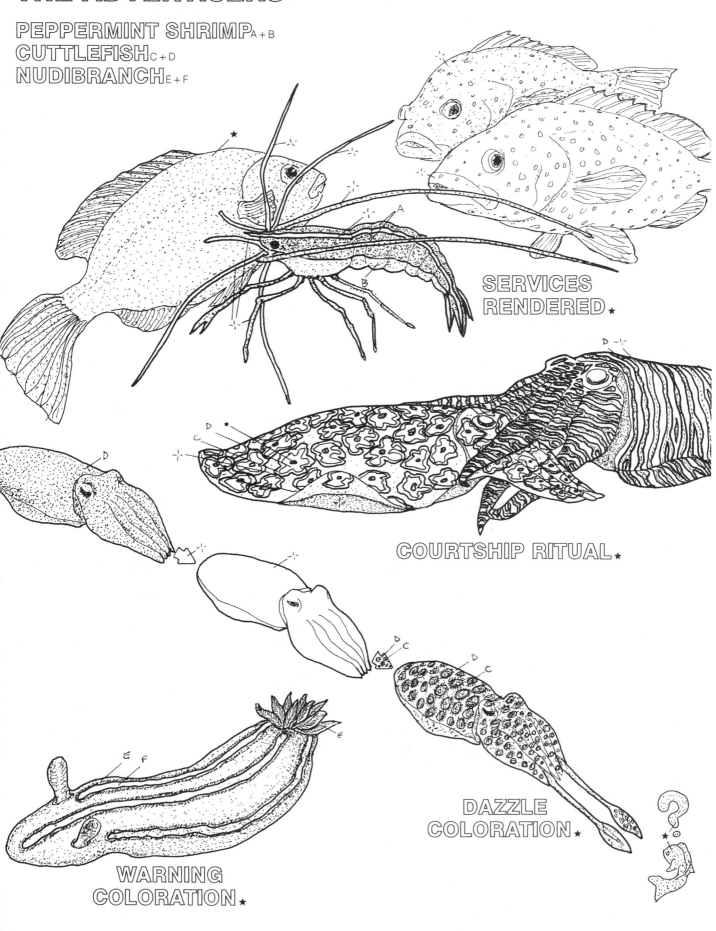

SERVICES
RENDERED ★

COURTSHIP RITUAL ★

DAZZLE
COLORATION ★

WARNING
COLORATION ★

53

COLORATION IN MARINE INVERTEBRATES: THE MIMICS

Many marine invertebrates use color as camouflage to mimic their environmental backgrounds as a protection against predation. As examples of such mimicry, this plate depicts four animals of the rocky intertidal, found along the Pacific coast of the United States.

In coloring this plate you may wish to use the natural colors of the invertebrates and their substrata as described in the text. As it is discussed, color each animal both in the large illustration and the smaller drawing in the center panel. Use red, green or brown for algae (F).

When seen against any other than its normal background, the red sponge *nudibranch* stands out vividly. However, this small (10 mm), bright red sea slug blends almost perfectly with the red *sponges* on which it lives, feeds, and lays its eggs. The *nudibranch* feeds by scraping away bits of the *sponge* with its radula (see Plate 89). The pigments of the *sponge* apparently are incorporated into the tissues of the *nudibranch,* resulting in the close color match between predator and prey. The red pigments are also deposited in the *nudibranch's* jellylike egg mass, which is laid on the *sponge* in a characteristic *egg spiral.*

Also shown on this plate is a particular species of small (15–20 mm) *isopod,* another invertebrate that mimics the color of the substrata on which it lives. This *isopod* feeds on both the green *surf grass* in the wave-swept lower intertidal, and on various *red algae* (see Plate 13) in the upper zones; in either case, its coloration matches the plant it is feeding on. Dr. Welton Lee, of the California Academy of Sciences, has studied these color changes and their significance. Adult *isopods* live and feed on surf grass, and they are green. When the adult females release the young *isopods* in this habitat they are unable to cling to the plants because of the strong wave surge. They are carried by the waves into higher tide zones where they cling to various *red algae.* Here the juvenile *isopods*

molt (shed their old outer skeleton) revealing a new, red skeleton below. In the living tissues beneath the reddish skeleton are pigment cells (chromatophores) (see Plate 54) which are used to adjust the color to the precise shade of red to match the particular *red alga* on which the animal is located. These animals feed on the *red algae,* grow, and upon reaching adulthood migrate back down to the *surf grass.* Once again they molt, and the new exoskeleton is green, matching their new substratum. This amazing color change ability allows the *adults* and *juveniles* to utilize different plants for food and attachment while both remain camouflaged as protection from predators.

The ribbed or digitate *limpet* is a common gastropod (see Plate 24) of the upper intertidal zone. Most of these *limpets* are found on bare rock and are generally dull brown to gray. However, another color variety is found in the middle intertidal zone associated with the common goose or *stalked barnacle* (see Plates 28, 81, and 95). These *limpets* have light gray shells with black stripes. When clinging to the *stalked barnacles,* the *limpet's* black stripes blend with the dark areas between the barnacle's gray plates. Thus camouflaged, the *limpets* are very difficult to detect, even when viewed at close quarters.

Several species of long-legged spider crabs "decorate" their bodies to blend with their backgrounds. The small (3–4 cm wide) *decorator crab* shown here lives among rock rubble in the low intertidal zones, or subtidally around kelp holdfasts. These habitats are often characterized by dense, low growths of attached plants and animals. The *decorator crab's* greenish-brown exoskeleton bears stout hooked "hairs," or setae, to which the animal attaches bits of local plants and animals (seaweeds, sponges, hydroids, and so on). These "decorations" grow and become living camouflage for the *crab.* When the *crab* remains motionless, it is extremely difficult to discern against the background of its habitat.

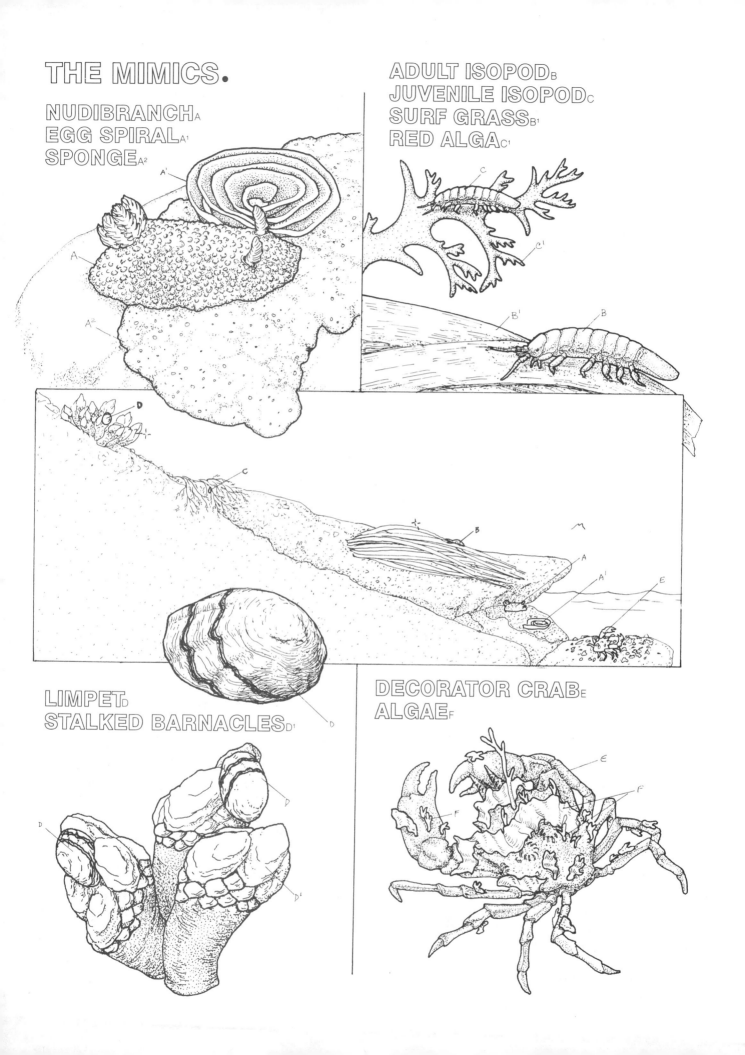

THE MIMICS.

NUDIBRANCH(A)
EGG SPIRAL(A1)
SPONGE(A2)

ADULT ISOPOD(B)
JUVENILE ISOPOD(C)
SURF GRASS(B1)
RED ALGA(C1)

LIMPET(D)
STALKED BARNACLES(D1)

DECORATOR CRAB(E)
ALGAE(F)

54
COLORATION IN MARINE INVERTEBRATES: CRUSTACEANS AND CEPHALOPODS

Marine invertebrates employ color changes both to advertise their presence and to mimic their surroundings. Crustaceans (crabs, shrimps, and others) and cephalopod molluscs (squids, octopuses, etc.) produce these color changes through the activities of special pigment-containing cells called *chromatophores*. However, the structure and methods of controlling the *chromatophores* differ greatly between these two groups of animals, as explained in this plate.

First read the text concerning crustacean chromatophores. Next, color the chromatophore diagram at the upper right in the plate, coloring only the pigment granules in their concentrated (E¹) and dispersed conditions (E²). Now color the pathway of the hormone (C) that controls the movement of pigment within the cell. Finish by coloring the left half of the fiddler crab as it appears in daylight (E²) with the same color used for the pigment granules (E¹).

The color of many crustaceans is often due, at least in part, to pigments deposited in the exoskeleton; these colors are relatively permanent. In some crustaceans, however, the exoskeleton is thin or translucent, and the colors of the *chromatophore* in the underlying tissue can be seen. A *chromatophore* is a single cell with highly branched processes radiating from its center. Each cell may contain one or several colors of *chromatophore pigment granules* (black, white, blue, yellow, red, or brown). When the *pigment granules* are concentrated in the center of the *chromatophore* they are more or less inconspicuous, but when the *granules* are dispersed and spread into the cell's branches, their color is plainly visible.

The movement of *pigment granules* within crustacean *chromatophores* is controlled by special *hormones* called chromotophorotropins. These *hormones* are usually secreted by special cells in the animal's *eyestalks* or *brain*. Apparently, each *pigment* color is individually regulated by particular and separate concentrating and dispersing *hormones*. Because these *hormones* must be carried through the blood from their sites of production to their sites of action, color change in crustaceans is relatively gradual.

Those crustaceans having a variety of colors of *pigment granules* can adapt to blend with nearly any background within a few hours. However, most crustacean color changes are much simpler, as in the example of the fiddler crab. This animal undergoes a daily color change involving very dark *pigment granules*. These *granules* are *dispersed* during the *daylight* hours, darkening the crab to blend with the muddy sand background of its normal habitat. At night, the *pigment granules* are *concentrated* and the crab appears pale in the moonlight.

After reading the text, color the pathway of nervous control of the cephalopod chromatophore (E). Next color the pigment granules (E¹) in the retracted, partially stretched, and completely stretched pigment sac (E³). Color only the active (stippled) nerves (F) and their associated contracted (shorter and thicker) muscles (G). Finally, color the concentrated and dispersed chromatophores (E) in the larval octopuses illustrated at the bottom of the page.

In contrast to the relatively slow-acting hormonally controlled *chromatophores* of crustaceans, those of cephalopods are regulated directly by the nervous system and can react in a fraction of a second. These fast-acting *chromatophores* are also very different structurally from those of crustaceans. The *pigment granules* are contained within an elastic *pigment sac*. *Muscles*, each with its own individual *nerve* for separate control, are attached to and radiate from the *pigment sac*. When some of these *muscles* contract, the *pigment* is partially dispersed by stretching the *pigment sac*. Complete *pigment* dispersion occurs when all of the *muscles* are stimulated to contract simultaneously, stretching the *sac* into a flat sheet of color. When the *nerve* stimulus ceases, the *muscles* relax, and the elastic *pigment sac* draws the *pigment* back into a condensed inconspicuous spot. The difference in color achieved between dispersed and condensed *chromatophores* is clearly seen on the small (2–3 mm) bodies of very young animals.

Adult cephalopods may have millions of variously colored *chromatophores* (yellow, orange, red, blue, black) arranged in patches and layers in their skins. By precise regulation of different *chromatophores* these animals can quickly change colors and color patterns to match their background, or flash various signals that play a part in their courtship, defense, and aggressive behaviors (see Plates 52, 65, and 87).

CRUSTACEANS.

EYE STALK_A
BRAIN_B
HORMONES_C
BLOOD STREAM_D
CHROMATOPHORE_E
PIGMENT
GRANULES_{E1}
DISPERSED (DAY)_{E2}
CONCENTRATED
(NIGHT)_{E-¦-}

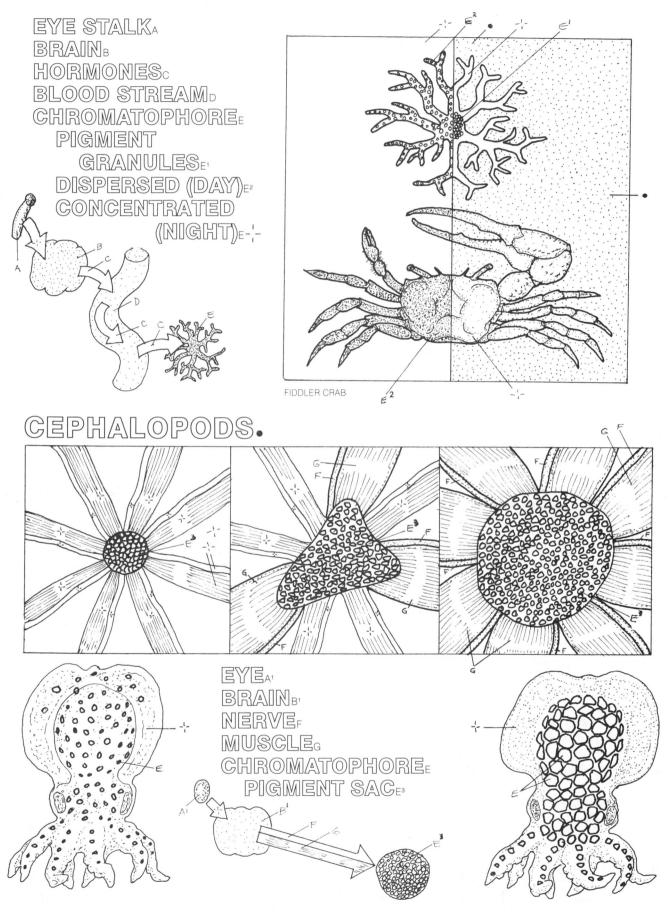

FIDDLER CRAB

CEPHALOPODS.

EYE_{A1}
BRAIN_{B1}
NERVE_F
MUSCLE_G
CHROMATOPHORE_E
PIGMENT SAC_{E3}

55
BIOLUMINESCENCE IN MARINE ORGANISMS

The ability to produce light is a characteristic of many marine organisms. The light is a product of an intracellular (within a cell) chemical reaction that requires oxygen, but which is still not completely understood. Some of the luminescent creatures of the sea are discussed below.

For accuracy and understanding you should use the described luminescent colors of the animals in this plate. Color each as it is treated in the text. Use blue-green for the luminescent wake (A) created by the boat. Then color the enlarged view of the planktonic organisms that create the glow. Use green to color the enlarged view of the fireworm (B) and the group of male worms (B²) swimming toward the ring (B¹) created by the female.

Many people who live near the ocean can recall dark summer nights when the sea appeared to glow with light. Each breaking wave had an eerie bluish shine, and boats left a luminous wake as they cut through the water. This commonly observed example of bioluminescence ("living light") is caused by tiny single-celled dinoflagellates, such as *Noctiluca*. When *Noctiluca* is agitated (as by a breaking wave or the propwash of a boat) it emits a brief blue-green flash of light. If present in large numbers, the combined luminescence of these organisms causes the glow observed at night.

As Columbus approached the coast of North America for the first time he reported seeing what appeared to be "candles moving in the sea." Some biologists suspect that Columbus was witnessing the mating ritual of the polychaete worm called the *Bermuda fireworm*. This small, bottom-dwelling worm swarms near the ocean surface on summer evenings for a few days following a full moon. The *females* swim in individual tight circles emitting a green *luminous secretion*. The *males* are attracted to this circle of light and swim toward it, emitting short bursts of bright light as they move through the water. Several *males* may be attracted to a single circle; soon both sexes release their gametes (sperm and eggs) into the water, where fertilization occurs.

Color the outline and the illustrated parts of the sea gooseberry (C, C¹, C²) red. Use bright blue for the group of small euphausiids (D) and the photophores (D¹) in the enlarged view.

The *sea gooseberry* (Phylum Ctenophora) is a marble-sized planktonic animal known also as a comb jelly or cat's eye. The illustration shows the animal's *comb rows,* which are bands of cilia used for locomotion, and the long *tentacles* used in capturing prey. The *sea gooseberry* is highly luminescent, and emits a fiery scarlet-red light when agitated. The light combines with the beating of the *comb rows* to create a truly beautiful and startling effect. The function of this animal's bioluminescent activity is not understood, but it may serve to confuse, frighten, or temporarily blind a potential predator.

Another bioluminescent zooplankter is the small (2–4 cm) crustacean known as a *euphausiid* (also called krill; see Plates 10 and 48). *Euphausiid* is a Greek word meaning "true shining light." The name refers to the blue bioluminescence produced by several distinct light organs *(photophores)* on the animal's body. These *photophores* tend to face downward (ventrally) and are located on the abdomen, thorax, and near the eyes. Many species of *euphausiids* occur in massive groups or schools that remain in deep, dimly lit water by day, and migrate into the upper layers at night. The luminescent activity of these crustaceans is thought to help keep the group together at depth during the daytime, and during their upward migrations at night. In some species, the production of light may function in mating behavior.

The photophores of the firefly squid (E) emit white light. You may wish to leave them blank and simply color around them in dark gray or black.

The *firefly squid* also employs bioluminescence in its mating behavior. In late spring, this small (10 cm), deep-sea *squid* migrates towards the surface to breed. This *squid* has a large number of *photophores* located on the mantle surface, around the eyes, and on the ventral arms. These *photophores* give off a brilliant white light that the *squids* flash during their nocturnal mating activities.

The *photophores* of most squids can be turned on or off and their intensity varied by direct nervous control. Some deep-sea squids that form schools probably use their bioluminescence to maintain contact with members of their species. Species that migrate to the surface at night to feed may employ *photophores* to produce "counterlighting," as described in the next plate.

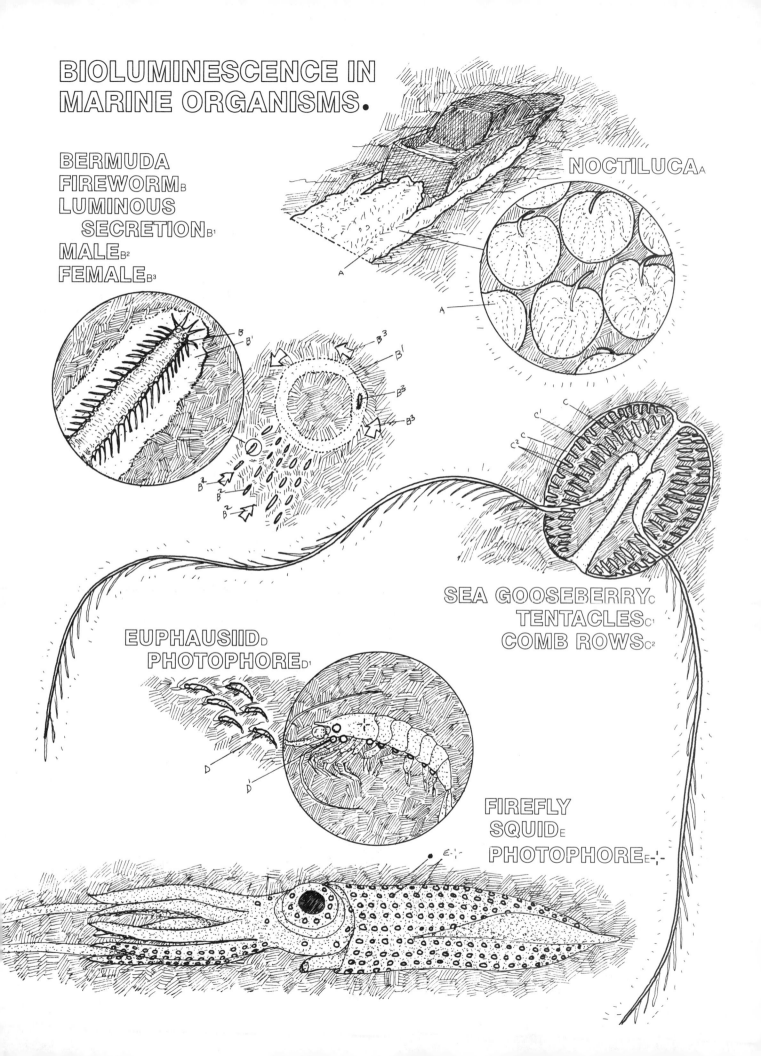

Sunlight penetrates only the upper layers of the ocean. In the dim mid-water realm and the black abyss, many fishes produce their own light. Like the squids and euphausiids discussed in the previous plate, some bony fishes have special *light*-producing structures called *photophores*. These structures are usually cup-shaped and may have elaborate focusing *lenses* and *reflectors* to concentrate and direct the *light* produced by the *photogenic cells*. These *photophores* are generally controlled by *nerves,* and are supplied with *blood vessels* that deliver the oxygen and energy sources needed for the *light*-producing chemical reaction.

Use a light blue to color the photophores (A), and yellow for the eye lens (H). The bars (A²) represent the bioluminescent glow. Color each fish as it is discussed in the text. Note that the bright flash (A²) of the caudal photophores of the lanternfish is represented as a burst of light. When coloring the flashlight fish, note that one is "blinking" and has its photophore (A³) covered by the skin fold (L).

Many mid-water fishes (see Plate 39) utilize bioluminescence for purposes that are often not fully understood, due to the difficulties in studying these animals alive. The *lanternfishes* have *photophores* along their ventral surface. At night these fishes migrate to the upper layers of water where they feed on zooplankton. Seen from below, an unlighted fish would be clearly silhouetted against the moonlit water surface, and thus be vulnerable to attack from below by predators using visual cues. *Lanternfishes* use their ventral *photophores* to erase this silhouette and blend with the moonlit upper waters. This phenomenon is known as counterlighting and occurs in a number of fishes, squids, and crustaceans (see Plate 55). The numbers and patterns of *photophores* vary between different species of *lanternfishes* and between sexes of the same species, thus providing mechanisms of recognition for schooling and mating.

The caudal (rear) *photophores* of many *lanternfishes* are used in evading predators. Seeing a threat, the fish emits a bright *flash* of light from the caudal *photophores* and, at the same time, swims rapidly away. The predator is startled and confused, and focuses its attention on the spot where the flash occurred, giving the *lanternfish* an opportunity to escape.

The *hatchetfish* uses its ventral *photophores* for protective counterlighting in the manner just described. It also possesses unique, upward-facing *eyes* equipped with *yellow lenses*. The *lenses* serve as filters that allow the fish to distinguish the narrow color range of bioluminescent light from the broad color range of normal background light. By looking up into moonlit water, the *hatchetfish* can discern potential prey animals that are using *photophores* in counterlighting behavior, thus actually capitalizing on the defense mechanism of its prey.

The deep-water predatory fish shown here is a *stomiatid.* It attracts potential prey with a lure equipped with a *photophore* and located on the fish's "chin whisker" or *barbel.* Of the species illustrated here, only *females* possess the *barbel. Males* are smaller and have a large *photophore* beneath their eyes. This *photophore* probably aids recognition between *males* and *females* in their dark environment.

The small (7–8 cm), shallow water, *flashlight fish* of the Red Sea bears *photophores* that are among the brightest and largest found in any bioluminescent organism. However, the blue-green light emitted from these *photophores* is not produced by the fish themselves, but rather by billions of luminescent bacteria harbored within the *photophore* by the host fish. The fish's blood stream keeps the microscopic inhabitants of this *bacterial photophore* supplied with nutrients and oxygen. The bacteria glow in a special pouch that is lined with dark skin arranged in such a fashion as to prevent the light from blinding the fish. As a control mechanism, the fish possesses a *fold* of *skin* that can be raised over the pouch to cover the *photophore* and essentially "turn off" the light. *Flashlight fish* remain hidden in the coral reef by day and on moonlit nights. On dark nights groups of from a few to sixty fish congregate near the surface. The combined glow of their *bacterial photophores* attracts their small zooplanktonic prey. If a larger potential predator is attracted to the light, the *flashlight fish* executes a strategic defense response known as "blink and run." The fish swims in one direction with its light "on," then covers the light and swims in a different direction. Each fish performs this "blink and run" behavior up to 75 times per minute. The effect of several *flashlight fish* "blinking and running" is to confuse the predator and permit the escape of the small fish.

BIOLUMINESCENCE IN MARINE FISHES.

MOON★

PHOTOPHOREA
LENS CELLB
PHOTOGENIC CELLA1 /LIGHTA2
REFLECTORC
NERVED
BLOOD VESSELE

LANTERNFISHF
CAUDAL FLASHA2

HATCHETFISHG
YELLOW EYE LENSH

MALE STOMIATIDI
FEMALEI
BARBELJ

FLASHLIGHT
FISHK
BACTERIAL
PHOTOPHOREA3
SKIN FOLDL

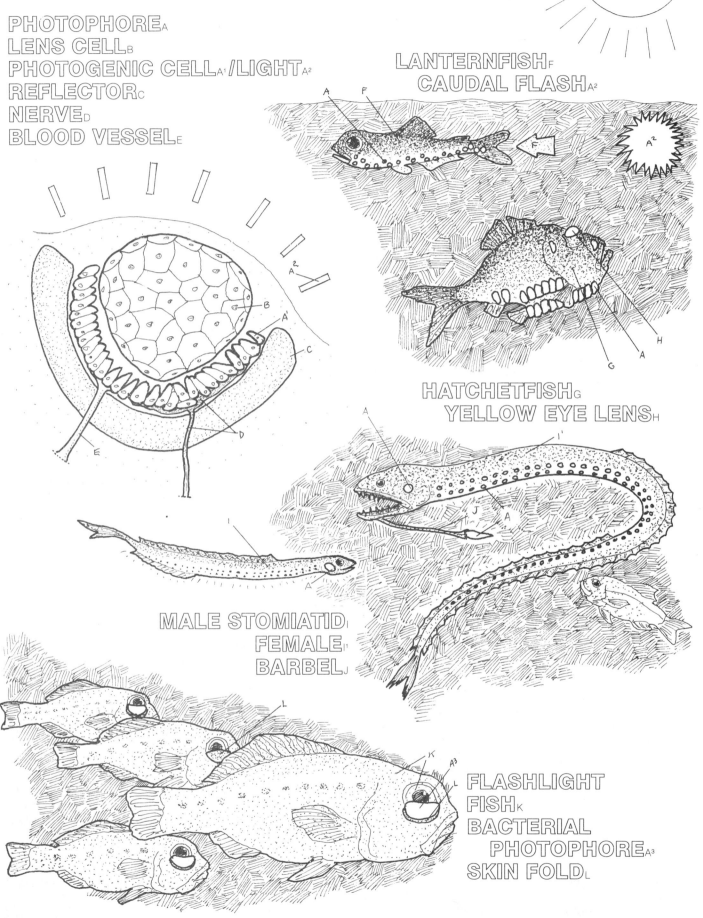

PHYTOPLANKTON REPRODUCTION: DIATOMS AND DINOFLAGELLATES

Diatoms and dinoflagellates are among the most abundant phytoplankters found in temperate coastal waters. In addition to reproducing sexually, both of these single-celled plants are able to take advantage of favorable conditions (sunlight, nutrients) by utilizing asexual reproduction to rapidly increase their numbers.

Before coloring, read the text concerning diatoms. Color the large diatom. Then color the diagrams of asexual reproduction. One complete asexual division is shown in the vertical column. The reduction in size with repeated asexual division is illustrated in abbreviated versions of the single process shown in the vertical column. Note that the offspring that are followed through subsequent divisions are labeled small offspring (B) and are stippled. The large offspring (C) also continue to divide asexually, but are not shown here. Note that the smallest diatom undergoes sexual reproduction.

A round, or centric, diatom typically found in the plankton consists of an outer, two-piece *frustule* composed of silica, surrounding the cell or protoplast (see Plate 12). To illustrate asexual reproduction, only the plates of the *frustule* and the cell *nucleus* are shown. The *nucleus* divides, and the interlocking pieces of the *frustule* separate. Each half of the original *frustule* serves as the outer piece of the *frustule* of the new individual. Thus, one of the offspring is the same size as the parent (the one formed from the original outer half of the parent's *frustule*). The other offspring, however, is *smaller* than the parent since its outer *frustule* plate is formed from the parent's smaller inner plate. In subsequent asexual divisions, descendants of the *smaller offspring* continue to decrease in size. When these asexually produced *offspring* are 60 to 80 percent smaller than the original parent, the *frustule* is too small to contain the necessary cell mass for proper functioning, and, under favorable conditions, sexual reproduction occurs. If environmental conditions are unfavorable at this time, the small diatoms will die.

Sexual reproduction in diatoms involves separate *male* and *female* individuals. The *frustule* of each cell separates, but its halves remain connected by a *membrane*. The *male* produces and releases flagellated, motile *sperm*, which swim to a *female* diatom containing a single *egg*. The membrane of the *female* separates allowing the entrance of *sperm* and subse-

quent fertilization of the *egg*. The fertilized *egg*, or *zygote*, swells to a globular shape, and the old *frustule* is shed. As the *zygote* develops, a new *frustule* is formed. The new *frustule* is large, the size of the original parent diatom, and the asexual cycle begins again. If there is sufficient sunlight, and the required nutrients are abundant, diatoms may undergo asexual reproduction more than once each day.

Color the illustrations of the asexual division of the theca (H²) and nucleus (D). Color the two types of cysts (I and J). Color the overlying water discolored by red tide (H¹) and the poison-containing shellfish (H⁴).

Dinoflagellates reproduce asexually by simple division of mature individuals. Shown here is an armored dinoflagellate that has a rigid cellulose *theca* (see Plate 12). The *nucleus* divides, and the *theca* separates. Each new *nucleus* remains in one of the halves of the *theca*, and a new second half is then formed. Unlike the diatom, both offspring grow to the same size as the parent. Under favorable conditions, dinoflagellates may divide every 8 to 12 hours. If conditions become unsuitable, the dinoflagellates form *temporary cysts,* which fall to the bottom. When conditions improve, the *temporary cysts* quickly become active cells again. Dinoflagellates are also capable of forming *resting cysts* by a sexual process (not shown) which is not fully understood. These *resting cysts* persist for months before becoming active again. *Cyst* formation in dinoflagellates allows these organisms to survive unfavorable environmental conditions.

Some species of dinoflagellates multiply to such incredibly high numbers (10–20 million cells per liter) that the water in which they occur becomes strikingly discolored. These *"red tides"* occur along all North American coasts and are triggered by a combination of biological and physical conditions. Certain dinoflagellates produce a toxic substance called *saxitoxin*. When mussels and clams feed on the dinoflagellates by filtering them from the water, this toxin becomes concentrated in their tissues. When bivalves (whether cooked or uncooked) containing high concentrations of this toxin are eaten by humans, a disease called paralytic shellfish poisoning results. In severe cases, the *saxitoxin* can cause death in 12 to 24 hours. Concentrations of *saxitoxin* sufficient to cause poisoning may be present in shellfish even when the *"red tide"* is not noticeable.

DIATOMS AND DINOFLAGELLATES.

DIATOM★
FRUSTULE★
   PARENTＡ
   SMALL OFFSPRINGＢ
   LARGE OFFSPRINGＣ
NUCLEUSＤ
MALEＥ
   SPERMＥ¹
FEMALEＦ
   EGGＦ¹
MEMBRANEＧ
ZYGOTEＥ¹＋Ｆ¹

ASEXUAL★

SEXUAL★

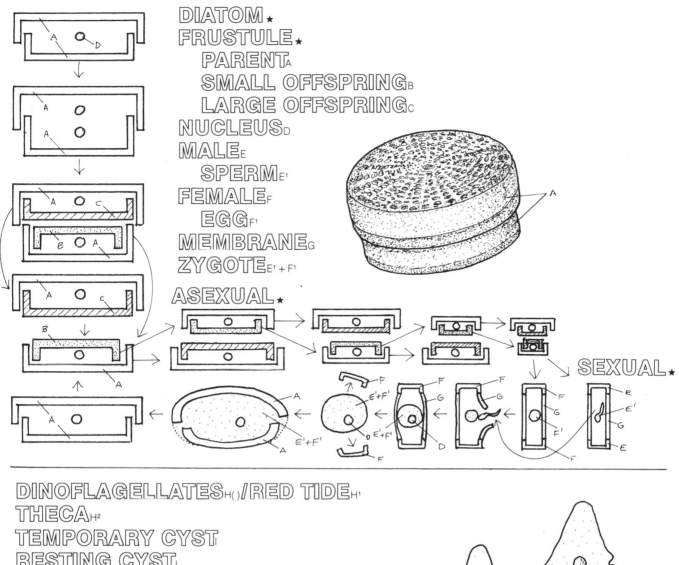

DINOFLAGELLATESＨ₍₎/**RED TIDE**Ｈ¹
THECAＨ²
TEMPORARY CYSTＩ
RESTING CYSTＪ
SAXITOXINＨ³
POISONOUS
   SHELLFISHＨ⁴

HOURSＩ

MONTHSＪ

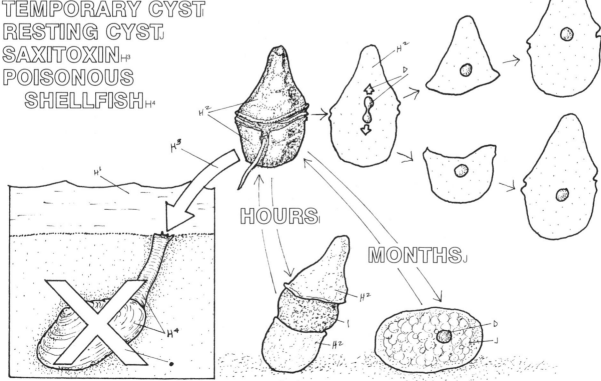

Much of the growth and proliferation of marine algae takes place by asexual or vegetative means. However, most algae have a sexual phase of their life cycle that is intimately involved in their overall survival strategy.

Begin with the diploid phase in the life cycle of *Ulva*. Locate and color the stages in each life cycle as they are mentioned in the text. The stages drawn within circles are microscopic in actual size.

The common green alga *Ulva* (sea lettuce) undergoes a cyclic alternation between a spore-producing plant *(sporophyte)* and a gamete-producing plant *(gametophyte)*. In *Ulva* these two types of plants appear identical without the aid of a microscope and detailed examination. Each cell of the *sporophyte* possesses a full complement of chromosomes (a condition designated as diploid). A cell division process called meiosis (reduction division) produces motile flagellated *zoospores,* each containing half the normal number of chromosomes (a condition designated as haploid). These haploid *zoospores* swim and attach to a hard substratum, where each can grow into a completely haploid *male* or *female gametophyte* plant. These plants then produce haploid *gametes* (*sperm* and *egg*) which are released into the water. A *sperm* and *egg* combine (fertilization) to produce a diploid *zygote* (two haploids = one diploid). The *zygote* settles and grows into another diploid *sporophyte* plant and the cycle is complete.

The large brown alga *Nereocystis* (bull kelp) also undergoes an alternation between diploid *sporophyte* and haploid *gametophyte*. In this case, however, there is a great difference in size between the two generations. The huge bull kelp with its 30-meter stipe is the *sporophyte* phase; it produces haploid flagellated *zoospores* in special regions of its blades. A single bull kelp might produce over 3.5 trillion *zoospores* in one year. The *zoospores* settle to the bottom and grow into separate, microscopic, male and female *gametophytes*. The male *gametophytes* produce motile *sperm,* which swim to the female *gametophytes* and fertilize the *eggs* in place, thus restoring the diploid chromosome number. The *zygote* grows as a *sporophyte* to eventually become the huge bull kelp.

Why the tremendous size difference between the *sporophyte* and *gametophyte* generations of this alga? Consider the timing involved. The bull kelp *(sporophyte)* grows rapidly during the spring and summer

months when optimal conditions prevail. It matures in late summer and produces *zoospores*. The microscopic *gametophytes* that grow from these *zoospores* produce their gametes and achieve fertilization during the winter, and the young *sporophytes* can usually be found in early spring. Many of the large *sporophyte* plants are torn loose from the bottom during fall and winter storms. The small *gametophytes* and young *sporophytes* are safe in cracks and crevices on the bottom during this inhospitable time, but are ready to take advantage of calmer weather and spring sunshine to start the cycle anew.

The red alga *Porphyra* is known as "nori" to the Japanese, who dry it and use it extensively in cooking. This seaweed represents a food industry involving $175 million annually. The details of sexual reproduction in this alga are not fully understood. However, a general knowledge of the life cycle is the basis for one of the most successfully managed marine cash crops in the world. (The raising and harvesting of marine organisms is called mariculture.) The stage of the life cycle that is dried and eaten is the *foliose* (leafy) *phase*. The *foliose phase* produces two types of *spores,* one that results in the growth of more *foliose phase* plants (the loop in the illustration), and one that generally settles on and bores into an empty mollusc *shell*. This shell-boring stage is known as the *conchocelis phase*. "Conch" means shell, and the name of this phase comes from *Conchocelis rosea,* a scientific name given to the shell-boring plant originally before its relationship to the *foliose phase* was understood. The *conchocelis phase* can also produce two types of spores. One type grows into more *conchocelis phases,* and the other type (called a *conchospore*) grows into the *foliose phase*. In nature, the *foliose phase* appears during the winter months in the high intertidal zone, where conditions are conducive to its growth. It is kept moist by rain and spray from storm waves, there is little competition from other plants, and the typical grazers of this area (limpets and littorines) are not particularly active. In the spring and summer the high intertidal is dry from exposure to the air and sun during daytime low tides, and grazing can be severe. *Porphyra* survives these adverse conditions in the *conchocelis phase,* safe in a shell on the bottom. The Japanese take advantage of the *conchocelis phase* by inducing it to release *conchospores* in closed tanks onto special ropes. The ropes are then suspended in the water to allow the *foliose phase* plants to grow and be harvested as nori.

LIFE CYCLES OF ALGAE.

SPOROPHYTE A
ZOOSPORE B
GAMETOPHYTE ★
MALE C
SPERM C¹
FEMALE D
EGG D¹
ZYGOTE C¹ + D¹
YOUNG SPOROPHYTE A¹

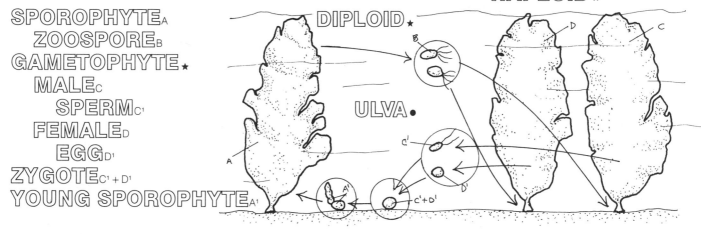

DIPLOID ★
HAPLOID ★
ULVA ●

NEREOCYSTIS ●
DIPLOID ★

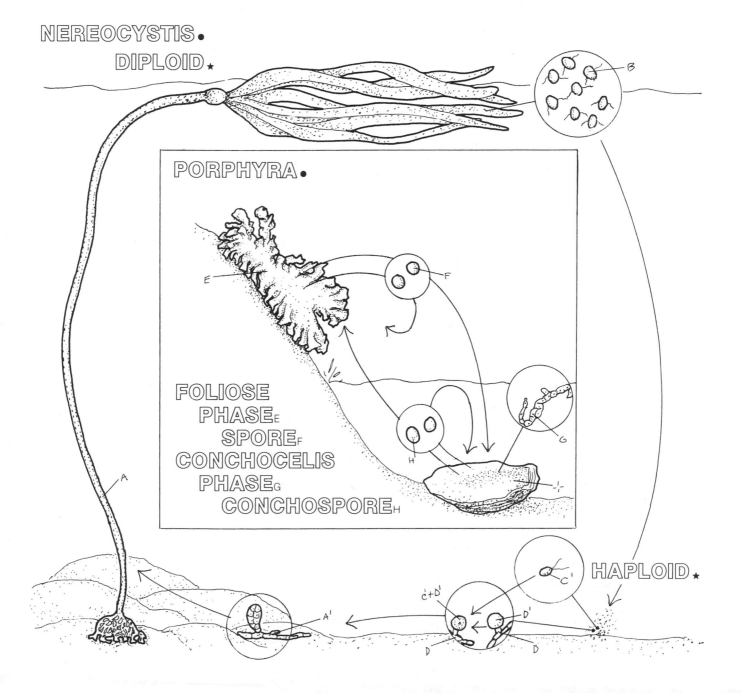

PORPHYRA ●

FOLIOSE
PHASE E
SPORE F
CONCHOCELIS
PHASE G
CONCHOSPORE H

HAPLOID ★

ASEXUAL REPRODUCTION

Sexual reproduction begins with the segregation of one-half of a cell's genetic material (chromosomes) into a sex cell, or gamete (sperm or egg). The union of sperm and egg (fertilization) restores the full complement of genetic material, creating a new individual with a combination of the genetic characteristics of the parents. Such genetic variation may result in organisms better adapted to their environments or that are able to adapt to new environmental conditions. This increased potential adaptability is one major survival advantage for species that utilize sexual reproduction. Some examples of these sexual processes are discussed in the following plates.

However, non-sexual (asexual) means of reproduction are used by many species to take advantage of favorable—and rapidly changing—environmental conditions. Great increases in numbers can be achieved more quickly by asexual than by sexual means. Five examples of asexual reproduction are presented in this plate.

Color each example of asexual reproduction as it is discussed. In the case of the sea star both illustrated processes involve regeneration, however the example on the right also results in the asexual formation of two individuals from one (reproduction).

In the previous plate it was explained how marine algae produce spores which asexually develop into new individuals. The single-celled diatoms and dinoflagellates (Plate 57) often reproduce asexually by simply dividing. The *turtle grass* shown here is an example of a marine flowering plant which produces new individuals by means of underground stems called *rhizomes* (see also Plate 11). Such *vegetative growth* also occurs as a type of asexual reproduction in many terrestrial plants.

Asexual reproduction is not always such a clearly definable phenomenon. For example, if a *sea star* loses all but one of its arms from the central disc it may gradually regrow the missing arms as shown. This process is called *regeneration*. However, if such a *sea star* is cut evenly in half (a generally futile practice of some oystermen who hope to eliminate the predators from oyster beds), each half might regrow the missing portion. Obviously this process involves *regeneration,* but it must also be considered *asexual reproduction* since two individuals are produced from an original one without sexual activity.

The coelenterates are masters of asexual reproduction. The large anemone *Metridium* (see Plate 16) exhibits a behavior known as *pedal laceration.* As the anemone creeps over the substratum on its pedal disc, portions of the disc are occasionally torn off and left behind, where they grow into small anemones. Some species, such as the aggregating anemone, reproduce by literally pulling themselves in half from top to bottom in the process of longitudinal fission (see Plates 3 and 79).

The ability to reproduce asexually allows a single organism to take advantage of a favorable habitat. For example, if a planktonic larval form finds a suitable site for attachment and growth, chances are excellent that others of its species would also be successful in that area. If the organism can reproduce by asexual means, it can proliferate and maximize its use of the habitat in a relatively short period of time. Such opportunistic exploitation of habitat is exemplified by corals. The coral's planula larva (see Plate 61) settles on an appropriate substratum and grows into a single *polyp.* If the *polyp* is successful, it doe not simply continue to increase in size, but instead *buds* off a new *polyp,* as illustrated. This *budding* process continues and can eventually produce large coral heads consisting of thousands of asexually produced individuals (see Plate 9).

A number of marine polychaetes (*Autolytus* is shown here) utilize a variant of asexual reproduction as a preparation for sexual reproduction. For most of the year, *Autolytus* exists in a sexually unripe, but otherwise adult form called an atoke. Preparatory to breeding, the posterior portions of the atoke begin to develop as a series of sexually maturing *epitokes,* complete with head (see illustration). These remain connected until the time of breeding; there is a record of one individual having twenty-nine *epitokes* attached. Breeding involves the synchronous release and swarming of multitudes of *epitokes* of both sexes, and this occurs, in most species, on one or a few nights each year. The timing of the event is precisely linked to the lunar cycle in a manner not yet understood. Swarming takes place in surface waters, with tremendous predation by birds, fish, and humans—some *epitokes* are much prized in Samoa, for example. This asexual multiplication of reproducing individuals, together with the simultaneous shedding of gametes, seems to maximize the chance of successful reproduction.

ASEXUAL REPRODUCTION.

VEGETATIVE
GROWTH A
TURTLE GRASS B
RHIZOME A1

REGENERATION C AND
ASEXUAL REPRODUCTION A2
SEA STAR D

PEDAL
LACERATION A3
METRIDIUM E

BUDDING A4
CORAL POLYP F

EPITOKE
FORMATION A5
AUTOLYTUS G

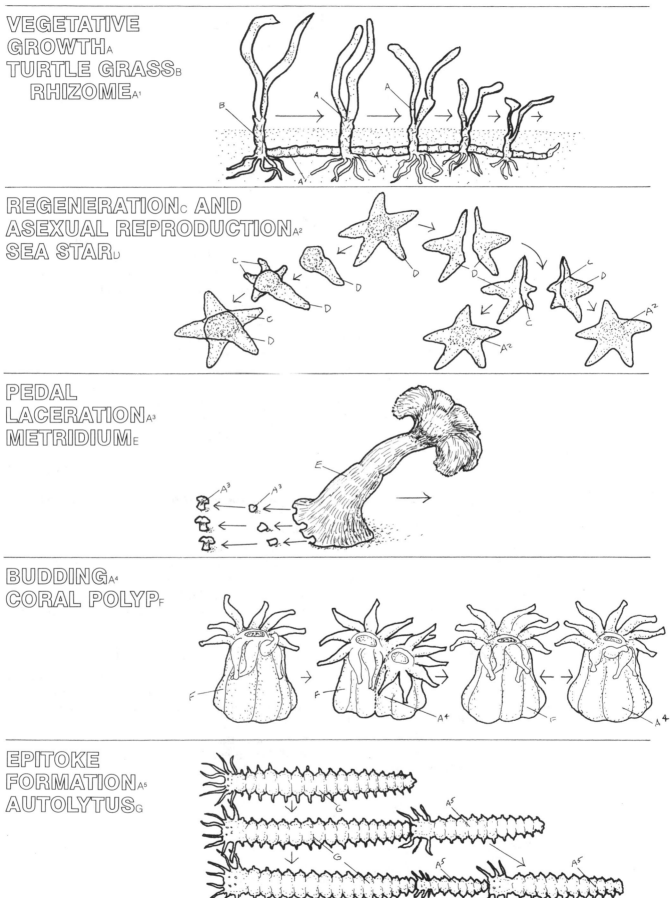

60
REPRODUCTION IN MARINE ORGANISMS: PLANKTONIC LARVAL FORMS

The embryonic development of many marine organisms includes a planktonic larval form. "Planktonic" indicates that the larvae float free of the bottom, and are incapable of swimming strongly enough to avoid being carried by ocean currents. The occurrence of a planktonic larval stage in the life cycles of so many different animals indicates that it must have tremendous advantages. First, the larva has an opportunity to feed on the richest of all marine food sources—the phytoplankton (see Plate 12). Second, ocean currents may transport the larva over great distances and allow dispersal to new habitats as well as replenishment of areas already occupied. Third, competition between adult and larva is virtually eliminated, as each exploits a different environment.

The production of planktonic larvae is not without expense. Many larvae never find their way to habitats that are suitable for adult life. Planktonic life is precarious. Microscopic drifting larvae may be subjected to areas of unfavorable water conditions, predators, and lack of available food. To compensate for the usually very high mortality rates among these planktonic stages, the adults produce enormous numbers (tens or hundreds of thousands per female) of small larvae. Also, the release of larvae into the plankton is generally timed to occur during or just prior to periods of high food availability (spring and summer). Such high production of larvae can sometimes backfire. If larval success is extremely high, the result may be overcrowding of the adult habitat after settling (for example, barnacles; see Plate 80).

As mentioned above, the habitats of larvae and adults are often totally different (for example, planktonic versus bottom-dwelling), and these two life stages usually consume different foods. Thus it is not surprising that the larval body form is entirely unlike that of the adult, as illustrated in this plate.

Color each adult/larva pair as it is treated in the text. Begin with the brittle star. All the larvae illustrated here are shown greatly enlarged. The setae (F) of the nectochaete larva are drawn with fine lines which may be lightly colored over. Only a simple outline and the eye of the adult sunfish are shown. Plate 36 includes a detailed drawing of this huge fish.

The larval form of the *brittle star*, or ophiuroids, (Phylum Echinodermata, see Plate 32) is the odd-looking *ophiopluteus*. It has eight *larval arms,* which are held somewhat rigidly by internal skeletal rods (not shown). The small (1 mm) *ophiopluteus* swims with the aid of tracts of cilia on the *arms* and body. It feeds on phytoplankton that are carried to the mouth by other ciliary tracts. The elongate *arms* help increase the surface area to volume ratio of the larval body, and offer resistance to sinking.

The *zoea larva* of the *porcelain crab* has elongated *larval spines* to help in flotation; the spines also serve as defense against predators. The adult *porcelain crab* bears little resemblance to its small (1.5 mm) *larval* form. The *larval spines* are lost during metamorphosis and the abdomen is tucked beneath the body—such extensions would hinder this much-flattened crab that hides in narrow places and under stones.

The tiny (0.25 mm) *nectochaete larva* of the polychaete *Nereis* (see Plates 20 and 62) shows the segmentation of the body so characteristic of the adult. Numerous spinelike *setae* on the larva's body offer resistance to sinking as this animal spends part of its life in the plankton. The *setae* also serve as paddles to propel the *larva* through the water.

Many fishes also have planktonic larval forms. The giant ocean *sunfish* has a tiny (3 mm) larva that looks very little like the adult. The pronounced spines develop as the larva grows (at about 13 mm) and are later lost as the adult form becomes apparent. These small larvae eventually become adults that grow to more than three meters and 2,000 pounds!

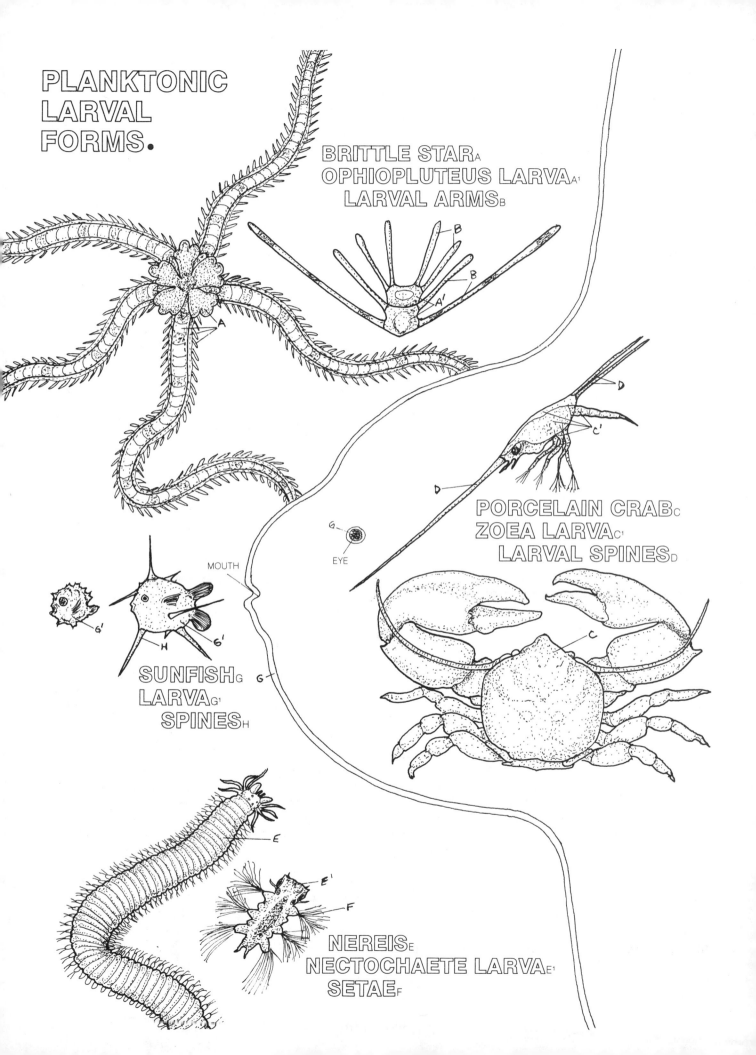

PLANKTONIC
LARVAL
FORMS.

BRITTLE STAR_A
OPHIOPLUTEUS LARVA_A1
LARVAL ARMS_B

PORCELAIN CRAB_C
ZOEA LARVA_C1
LARVAL SPINES_D

MOUTH

EYE

SUNFISH_G
LARVA_G1
SPINES_H

NEREIS_E
NECTOCHAETE LARVA_E1
SETAE_F

61

REPRODUCTION IN COELENTERATES: COELENTERATE LIFE CYCLES

Coelenterates exhibit two basic body forms or morphs, the polyp and the medusa (see Plates 16 and 17). In general, polyps are sessile and reproduce by asexual methods (budding), while the medusae are free-floating and reproduce by sexual means (there are many exceptions and modifications to this basic plan). Both morphs occur in the life cycles of some coelenterates, while others lack one or the other form. In addition, most coelenterate life cycles involve both sexual and asexual reproduction. This plate discusses the life cycles of a representative of each of the classes of the phylum Coelenterata.

Begin with the life cycle of *Obelia*. First color the animal colony on the left, then follow the arrows through the cycle as it is discussed. The medusae are exaggerated in size, as are the sperm, eggs, and planula larva.

The feeding polyp stage of the class Hydrozoa (here represented by *Obelia*) is called a *hydranth*. These *hydranths* occur singly in some species, but more commonly as *colonies* of asexually produced individuals connected by a common gut tube, or coelenteron (see Plate 16), and branching *stalk*. An *Obelia* colony resembles a small bushy alga to the naked eye, but microscopic examination reveals the tiny (0.2 mm) *hydranths* along the colony's branches. At certain times of the year one can find reproductive polyps or *gonozooids* interspersed among the *hydranths*. Visible in the illustration are numerous *medusa buds* forming on the *gonozooids*. These are produced asexually and released as tiny free-swimming *medusae*. Each *medusa* is either male or female, and as they mature each produces either *sperm* or *eggs,* which unite to form a *zygote* (fertilized egg). The *zygote* develops into a free-swimming *planula* larva, which eventually settles on a firm substratum and becomes a small *hydranth* that begins a new colony by asexual budding.

Begin coloring the *Aurelia* life cycle with the male and female adult medusae, and continue as each stage is mentioned.

In the common jellyfish *Aurelia* (Class Scyphozoa) the *medusa* is the largest and dominant stage of the life cycle. *Medusae* 10 to 20 centimeters in diameter often aggregate in huge numbers in the coastal waters of North America. The sexes are separate in *Aurelia*, and the female broods the *zygotes* on her *oral arms* until they reach a free-swimming *planula* stage. The *planula* swims for a short while, until settling on a solid substratum, whereupon it grows into the polyp stage called a *scyphistoma*. The *scyphistoma* feeds on small zooplankton in typical polyp fashion (see Plate 16) for some time and then begins to partition its body into a stack of tiny potential *medusae*. This process is called strobilation, and the polyp is now called a *strobila*. One at a time, these asexually produced (budded) young *medusae* are released, swim away, and mature to begin the cycle once again.

Color the life cycle of *Epiactis*. The arrow (E + F) in the sea anemone's (*Epiactis*) life cycle indicates the transfer of the zygotes (E + F) from the female's mouth (K) to the base of her column (M).

Sea anemones and corals (Class Anthozoa) have only the polyp stage in their life cycle, but most are still capable of both sexual and asexual reproduction. The small sea anemone illustrated here, *Epiactis,* is a common species on the west coast of the United States, and has been subject to some past misunderstanding. The adults of this anemone are often found with many small *juveniles* attached to the base of their *columns*. Since anemones commonly reproduce asexually, it was assumed that these young individuals were the result of budding by the adult, hence the vernacular name "proliferating anemone" (*prolifera*). The faulty assumption was exposed some years ago by Dr. Daphne Dunn, who studied these anemones while at the University of California, Berkeley. Based on her work, it is now known that adult *Epiactis* retain their fertilized eggs *(zygotes)* within their gut cavity (see Plate 16) where they are brooded for some time. The embryos eventually emerge from the anemone's mouth, crawl to the base of the *column,* and attach. Here the *juveniles* grow to about 4 millimeters in basal diameter, before crawling away to establish their own independent adult life.

COELENTERATE LIFE CYCLES.

OBELIA ★

COLONY ★
  STALK A
  HYDRANTH B
  GONOZOOID C
    MEDUSA BUD D
MEDUSA D'
  SPERM E
  EGG F
  ZYGOTE E I F
PLANULA G

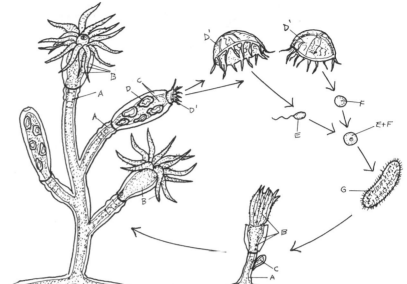

AURELIA ★

MEDUSA D'
  ORAL ARM H
  SPERM E
  ZYGOTE E+F
PLANULA G
SCYPHISTOMA I
STROBILA I'

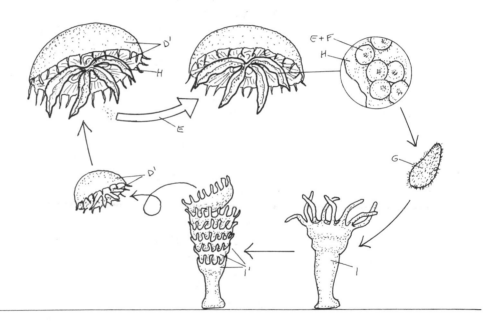

EPIACTIS ★

ADULT ★
  ORAL DISC J
  MOUTH K
  TENTACLES L
  COLUMN M
SPERM E
ZYGOTE E+F
JUVENILE N

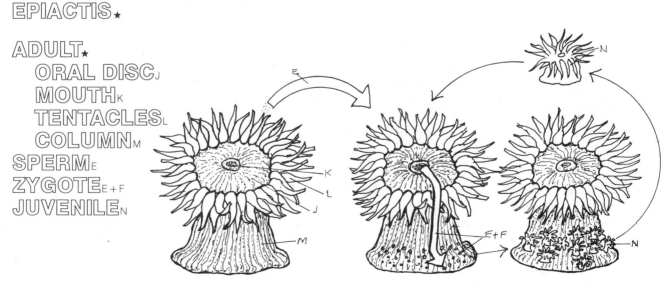

62

REPRODUCTION IN MARINE WORMS: POLYCHAETE LIFE CYCLES

The reproductive cycles of the polychaete worms (and many other animals) are tightly coupled to the phases of the moon. The exact mechanism of this correlation has yet to be discovered. This plate introduces the reproductive behavior of three polychaete worms and the role of the moon in each life cycle.

Begin with the small, tube-dwelling worm *Spirorbis* in the center of the page. The enlarged illustration shows the worm in its tube; portions in which the male (B) and female (C) segments lie are labeled. The newly settled larvae (E) and adult worms are shown on a blade of oar weed which is not to be colored.

The small (1–2 mm diameter), spiral, calcareous (lime) tube of the polychaete *Spirorbis* is a common sight on many marine substrata. These tiny filter-feeding worms settle on rocks, mollusc shells, and seaweeds. Unlike most polychaetes, in which the sexes are separate, *Spirorbis* is a hermaphrodite. The anterior segments of the worm contain *female* sex organs, and the posterior segments contain *male* organs.

Many *Spirorbis* species follow a reproductive cycle that is closely tied to the moon's phases and occurs at the same lunar phase each month. Fertilized eggs of this worm are retained in the *tube* of the adult, unlike most other polychaetes which release their eggs and sperm into the water. Each month free-swimming *larvae* are released to seek out new habitats. Settling behavior of a *larva* involves landing on a substratum to "feel" and "taste" it. If it is judged unsuitable, the larva swims off to settle and try again. Different species of *Spirorbis* prefer different types of substrata. The *tube*-encrusted blade of the oar weed (see Plate 14) shown here illustrates two aspects of settling behavior in *Spirorbis*. The individual worm *tubes* are evenly spaced. A settling *larva* crawls back and forth over the area and, presumably through some avoidance mechanism, will remain only if sufficient room is available to allow its growth to adult size. Note also that the *larvae* have settled on the innermost part of the blade, nearest the stipe. This is the point from which new blade growth arises. By settling on this area of new growth, *Spirorbis* is least likely to encounter other attached organisms, and, since the blade wears back from the tip, the innermost part offers the longest-lasting habitat.

Color the worm *Nereis*, in the lower right hand corner. Color the enlargement showing the crawling setae (H), and color them along the worm as well. Then color the heteronereid (I) which is shown swimming at the surface using its modified swimming setae (J).

Many benthic polychaetes (Plate 20) spawn at the surface, rather than in their normal habitat. In *Nereis* the bottom-dwelling adults undergo a radical transition in morphology (body appearance) into the sexual *heteronereid* form. The *eyes* become enlarged, and the appendages on the posterior segment change from a crawling to a swimming function. The pointed *crawling setae* are discarded and paddle-shaped *swimming setae* take their place. The parapodia also develop elaborate chemosensory (chemical sensing) organs that alert the worms to the presence of other *heteronereids*. The *heteronereids* mature and become packed with ripe sex cells (gametes). At precise phases of the moon, the *heteronereids* swim to the surface (shown in the upper illustration). The males release their sperm in controlled amounts, stimulating the females to rupture (split open) and release their eggs. Fertilization takes place in the water.

Color the palolo worm and its stolon in the lower left corner. Then follow the arrows to the stolons swarming at the ocean's surface. The swimming heteronereid is drawn in an exaggerated scale relative to the palolo stolon.

To insure that individuals assemble for reproduction, a behavior called swarming occurs in many species. The best-known example of polychaete swarming involves the Samoan *palolo worm*. This polychaete lives in crevices of shallow coral reefs of Samoa, Fiji, and other South Pacific islands. The adult *palolo worms* grow an elongated posterior structure, called the *stolon*. The *stolon* segments are narrow and long and each has a light-sensitive ventral (bottom) *eyespot*. On the eighth and ninth days after full moon in October and November, the adult *palolo worm* backs out of its crevice and releases the *stolon*. The *stolons* swim to the surface in vast swarms, rupture, and release their sex cells. Fertilized eggs may cover acres of the sea surface. The island natives consider the *palolo worm* a delicacy and await the pre-dawn swarm with dip nets and hearty appetites.

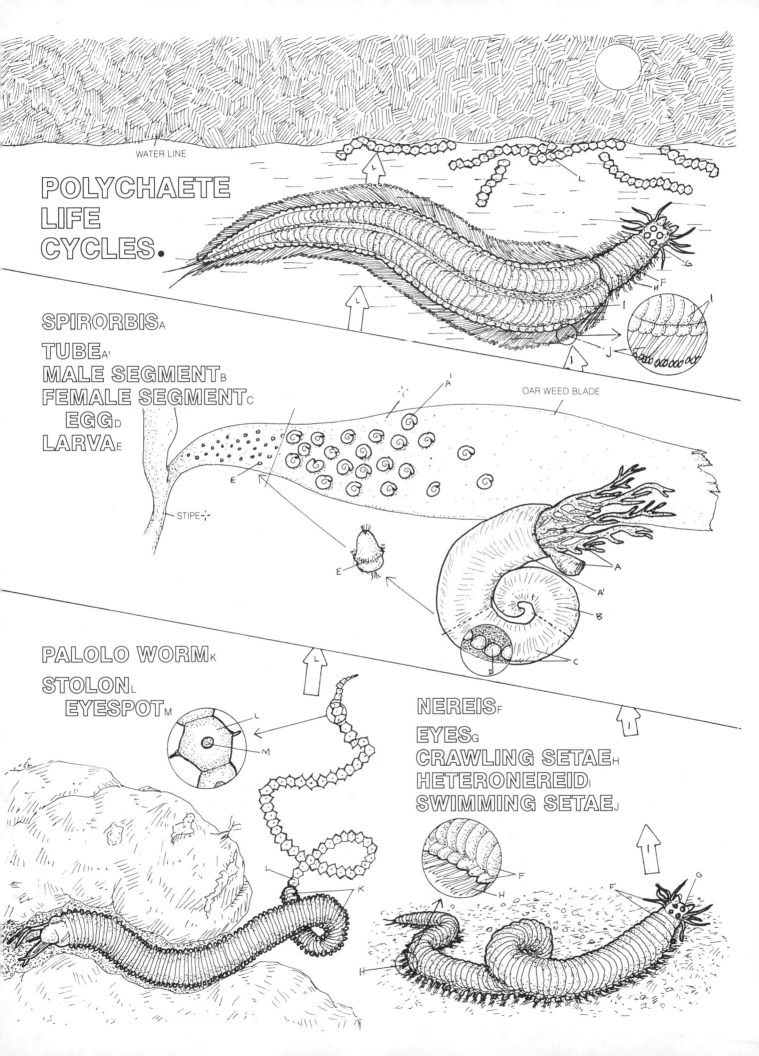

WATER LINE

POLYCHAETE LIFE CYCLES.

SPIRORBIS A

TUBE A¹
MALE SEGMENT B
FEMALE SEGMENT C
EGG D
LARVA E

STIPE

OAR WEED BLADE

PALOLO WORM K

STOLON L
EYESPOT M

NEREIS F

EYES G
CRAWLING SETAE H
HETERONEREID I
SWIMMING SETAE J

Most species of bivalves are dioecious (sexes are separate) and reproduce by releasing their sex cells into the sea water (spawning). *Fertilization* occurs in the water, and a planktonic *larva* develops (see Plate 60). This plate illustrates and explains the life cycle of the Virginia Oyster, and some aspects of the commercial importance of this animal.

Color the spawning male (A) and female (B) oysters and their sperm (A¹) and eggs (B¹). Note that fertilization (A¹ + B¹) takes place in the overlying water. Follow the development of the fertilized eggs and color each stage as it is mentioned in the text. Note that the veliger larva settles on an empty shell. New shell growth of the young oyster receives the same color as the settled spat (I).

Since oysters tend to settle in aggregations, the *males* and *females* are in close proximity and need only spawn into the water at the same time to insure *fertilization*. The fertilized eggs undergo *cell division* and develop in the water column. A ciliated *trochophore larva* develops which then becomes a *veliger larva* bearing a rudimentary hinged bivalve *shell,* and a ciliated *velum* for swimming. The *veliger larva* feeds on *phytoplankton* (see Plate 12) and continues to grow. After two to four weeks, a *foot* develops, and the *veliger* begins contacting the bottom with its extended *foot.* When a solid substratum is encountered, the *veliger* crawls about, testing the bottom. If the substratum is unsuitable (texture, crowding, or other factors) the *veliger* swims off to try again. When an appropriate substratum, such as an *empty shell,* is found, the *veliger* produces a cement (from pedal glands associated with the *foot*) and attaches its left valve to the substratum. Once secure, the *velum* is lost, the *foot* becomes very reduced, and the animal begins to grow. This newly attached oyster is called a *spat.* Its gills develop (not shown) and it begins its life as an attached, filter-feeding organism (see Plate 22). As the *shell* enlarges it becomes irregularly shaped, as

shown in the diagram of the two adults.

Virginia oysters are able to withstand reduced salinities and frequent exposure to air in the intertidal zone. Thus they can grow in estuaries, where rivers meet the sea and form quiet embayments and lagoons with broad mud and sand flats (see Plate 5). Oyster *larvae* seek out and settle on hard substrata in these habitats, often in tremendous numbers. Their preference for settling on the shells of adult oysters (or other bivalves) may result in the development over time of thick oyster "beds." This situation was recognized very early, and elaborate oyster cultivation practices date back to the Roman Empire. Today oyster culture is big business, and it is an understanding of the natural life cycle and settling behavior that form the basis for these endeavors.

Color the scallop shells to which the oyster spat (I) are attached. After reading the text, complete the coloring of the oyster culture raft.

Oyster culturists maintain mature *male* and *female* oysters in aquaria and induce them to spawn by a sudden elevation of the water temperature. The fertilized eggs are collected and placed in tanks, where they are raised to *veligers* on a rich diet of *phytoplankton.* The maturing *veligers* are then transferred to other aquaria containing empty *scallop shells* to which the oyster *spat* attach. These *scallop shells* with their attached oysters are then moved out to the oyster beds. The oysters grow and can reach market size in from two to five years, at which time they are harvested.

A more productive (more oysters per unit of available area) method of oyster culture is called *raft* culture. In this case, the *scallop shells* with *spat* attached are strung on *ropes* and suspended from a floating *raft.* By this arrangement the oysters are protected from benthic predators, such as sea stars, and spared the stress of exposure to air at low tide. In waters rich in *phytoplankton,* these oysters feed continuously and grow rapidly.

BIVALVE REPRODUCTION: VIRGINIA OYSTER.

MALE_A
   SPERM_{A¹}
FEMALE_B
   EGG_{B¹}
FERTILIZATION_{A¹ + B¹}
CELL DIVISION_C
PHYTOPLANKTON_D

TROCHOPHORE LARVA_E
VELIGER LARVA★
   SHELL_F
   VELUM_G
   FOOT_H
SPAT_I
EMPTY SHELL_J

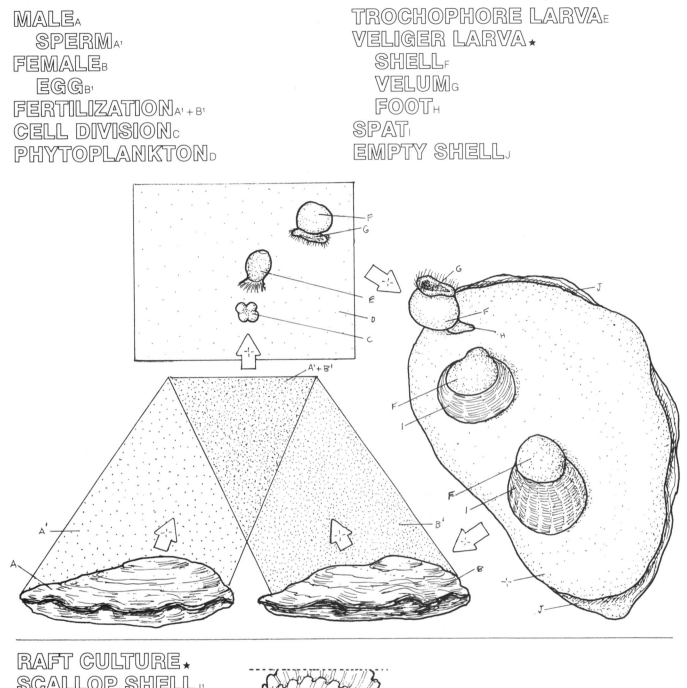

RAFT CULTURE★
SCALLOP SHELL_{J¹}
ROPE_K
RAFT_L

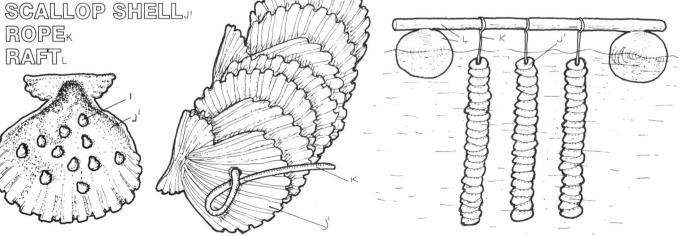

64
REPRODUCTION IN MOLLUSCS: GASTROPOD LIFE CYCLES

Gastropods are the most diverse group of molluscs. They include large herbivorous (plant-eating) abalones, shell-less nudibranchs, or sea slugs, the familiar garden snails and slugs, and many other forms. To illustrate this great variety, the life cycles of four representative gastropods are presented in this plate.

Color each life cycle as it is mentioned in the text. Beginning with the abalone, note that the large excurrent openings of each shell receive the color of the sperm or eggs that are released through them. The zygote (B + D), receives the color of the sperm (B) and eggs (D); also color the title "zygote" with both colors. The size of the trochophore (E) and veliger (F) larvae are greatly exaggerated here.

The life cycle of the abalone is fairly similar to that of the bivalve mollusc reviewed in the previous plate. The sexes are separate, and *sperm* and *eggs* are released into the water through excurrent openings in their shells. The fertilized egg, or *zygote,* develops in the water, first into a ciliated *trochophore* larva, and later into a *veliger* larva. The gastropod *veliger* differs from the bivalve veliger, among other ways, in forming a typical, snail-like, coiled shell. After several weeks in the plankton, the *veliger* resorbs its ciliated *velum* and settles onto a rocky surface as a *juvenile.* In the abalone, settling is triggered by a chemical given off by certain algae, thus assuring a food supply for the young gastropods. The abalone becomes sexually mature in two to five years, depending on the species.

Color the moon snail and the whelk mating aggregation. The female moon snail is shown producing a sand collar (D¹) after mating with the male.

The life cycle of the moon snail (see Plate 24) is somewhat more complex than that of the abalone. The sexes are separate, but the moon snails do not spawn into the water. The *male* possesses a penis with which he transfers *sperm* into the *female,* where the eggs are fertilized. The *female* lays the fertilized eggs in a case made of mucus and sand, called a *sand collar* because of its resemblance to an old-fashioned celluloid collar. The embryos develop within the sand and mucus

matrix. The *sand collar* eventually deteriorates, releasing the moon snail *veliger* larvae into the water. The *veligers* grow and seek out a fine sand bottom on which to settle and assume life as *juveniles.*

The whelk is a relatively advanced shelled gastropod, with separate sexes and internal fertilization. Reproduction is generally seasonal, and it is not unusual to find several dozen whelks clumped together in a *mating aggregation* in the rocky intertidal zone. The *females* lay their fertilized eggs in small (5 mm), yellowish, vase-shaped *egg capsules* which are attached to rocks shielded from direct sunlight. The eggs pass through a modified development within the *capsule;* no free-swimming larval stages are formed. As the young snails develop within their *capsules* they become cannibalistic, eating one another until only a single (well-fed) *juvenile* emerges from each *capsule.* This direct, localized development and release of young insures a proper habitat for the new snails, but relinquishes the benefits enjoyed by those animals with planktonic larvae (see Plate 60).

Color the dorid nudibranchs. The adult nudibranchs are given both male and female colors since they have both reproductive systems and are pictured fertilizing each other.

The sea slugs, or nudibranchs (Plate 25), have a life cycle that combines some aspects of the patterns explained above with a few variations of their own. Most nudibranchs are predators, grazing on various hydroids, sponges, and so on, and are generally solitary in their habits. Nudibranchs are hermaphroditic (possess both male and female sex organs), and the meeting of two adults for reproduction involves the simultaneous exchange of *sperm* with the subsequent fertilization of each animal's eggs. Thus their solitary habits are somewhat countered by this double mating that results in two animals carrying fertilized eggs. Each nudibranch species lays its eggs in a characteristic egg cluster; shown here is the *egg spiral* of a dorid nudibranch. Other species lay their eggs in strings or flat sheets. After developing within the egg cluster, *veliger* larvae are released, each complete with a rudimentary coiled shell. This shell is lost when the larva settles and life as a *juvenile* begins.

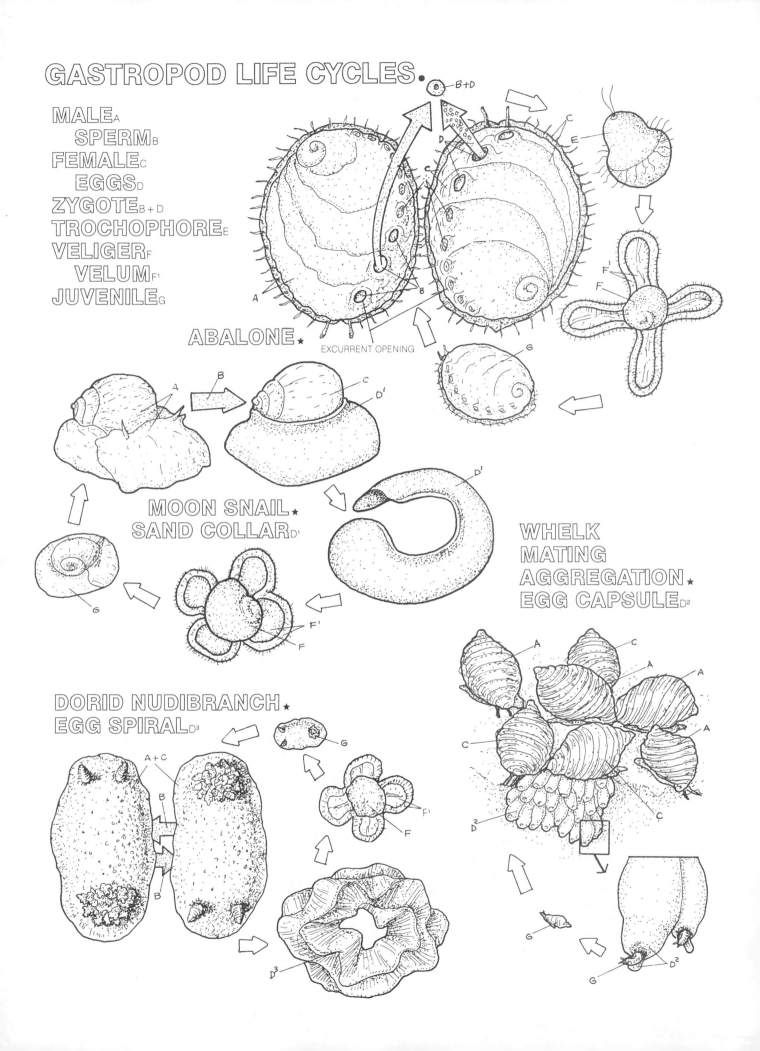

GASTROPOD LIFE CYCLES.

MALE$_A$
SPERM$_B$
FEMALE$_C$
EGGS$_D$
ZYGOTE$_{B+D}$
TROCHOPHORE$_E$
VELIGER$_F$
VELUM$_{F^1}$
JUVENILE$_G$

ABALONE ★

EXCURRENT OPENING

MOON SNAIL ★
SAND COLLAR$_{D^1}$

WHELK
MATING
AGGREGATION ★
EGG CAPSULE$_{D^2}$

DORID NUDIBRANCH ★
EGG SPIRAL$_{D^3}$

65

REPRODUCTION IN MOLLUSCS: CEPHALOPOD REPRODUCTION

The reproductive behavior of cephalopods is quite complex and involves some unique structures. Three species are discussed here to illustrate some of these features.

Begin by coloring the mating octopuses and proceed down the page as each species is discussed. Note that the male (A) cephalopod has a modified, hectocotylized arm (B) which receives a different color from the rest of the body. In the octopus, the small (2–3 cm) spermatophore (C) is exaggerated in size as are the juvenile (G) octopus and squid. In the paper nautilus note that the female's membrane (H) which secretes the shell (I), receives a separate color. The diagram at lower left illustrates a shell with the egg cluster attached.

The common shallow-water octopus of Europe and the east coast of the United States is a relatively hardy species. Because it does well in captivity, more is known of this cephalopod's reproductive processes than of many other species. This normally solitary animal displays little in the way of a courtship ritual. A *male* approaches a *female* and attempts to mate. He may actually mount the *female* or simply extend the third arm on his right side toward her, as shown. This arm is the *hectocotylized arm;* its tip is modified into a spoon-shaped palm called the *hectocotylus* (hecto = hundred; cotyl = cup) that is used to transfer sperm into the *female's* mantle cavity. The *hectocotylized arm* bears a groove along its posterior surface. The sperm are packaged in gelatinous sheaths known as *spermatophores* which are loaded into the sperm groove by the penis (not shown). Waves of muscular contraction along the arm carry the *spermatophores* along the groove to the *hectocotylus,* which is inserted into the *female's* mantle cavity, near her reproductive opening. (In some species, the *hectocotylus* is detached within the *female's* mantle cavity and remains there. Some early investigators mistook this structure as a parasitic worm and classified it as the sole species of the genus *Hectocotylus.* Although the genus has since been declassified, the name *hectocotylus* for the modified arm or its tip has been retained.) The *spermatophores* rupture inside the *female's* reproductive tract, assuring that the sperm reach their destination. A *male* may transfer as many as fifty *spermatophores* in an hour. After mating, the *female* leaves to lay her eggs, a process which may take a week as she delicately releases a few eggs at a time through her *funnel.* She positions thousands of eggs in grapelike *egg clusters* suspended from the rocky walls of her cave home. The *female* remains with her eggs for several weeks, constantly manipulating them and gently blowing water over them with her *funnel* to keep them clean and oxygenated. The *juvenile* octopuses hatch as miniature adults and spend weeks in the plankton before settling on the bottom. The *female* does not eat during her care of the *eggs,* and dies soon after hatching of her young.

The common Pacific squid congregate and mate in large groups. As a *male* approaches a prospective *female* his arms and head are ablaze with red and white markings which change to generally solid maroon when actual copulation ensues. The animals mate either head on (see Plate 54) or with the *male* grasping the *female* from under the left side, as shown here. Using the lower left *hectocotylized arm,* the *male* picks up spermatophores delivered to his funnel by the penis (not shown), and transfers them into the *female's* mantle cavity. He holds them within the mantle cavity until the sperm are released and then withdraws. The entire copulation process may take less than ten seconds. Following mating, the *female* produces long *cases* containing the fertilized eggs. The *egg cases* swell and become firm when exposed to sea water. They are attached to the bottom, often in large clusters or "mops," covering many square feet. After several bouts of mating and egg laying, the adult squids die. The young squids develop within the *egg cases* and eventually emerge as *juveniles,* complete with functional chromatophores and ink sacs (see Plates 54 and 87).

The *female* pelagic octopus, known as the paper nautilus, prepares an elaborate ark for her eggs. Each of the *female's* two dorsal arms bears a broad *membrane.* Each *membrane* secretes half of a *shell* in which the *female* attaches her *egg cluster.* The *female* remains in the *shell* with only her arms and funnel protruding. The smaller *male* often occupies the *shell* with the *female* and her *egg cluster.* This octopus is called the paper nautilus because the thin *shell* superficially resembles that of the nautilus (see Plate 26), but the two cephalopods are not very closely related.

CEPHALOPOD REPRODUCTION.

OCTOPUS ★
MALE_A
   HECTOCOTYLIZED ARM_B
      HECTOCOTYLUS_B'
   SPERMATOPHORE_C
FEMALE_D
   EGG CLUSTER_E
FUNNEL_F
JUVENILE_G

SQUID ★
MALE_A
   HECTOCOTYLIZED ARM_B
FEMALE_D
   EGG CASE_E'
JUVENILE_G

PAPER NAUTILUS ★
MALE_A
   HECTOCOTYLIZED ARM_B
FEMALE_D
   MEMBRANE_H
SHELL_I
EGG CLUSTER_E

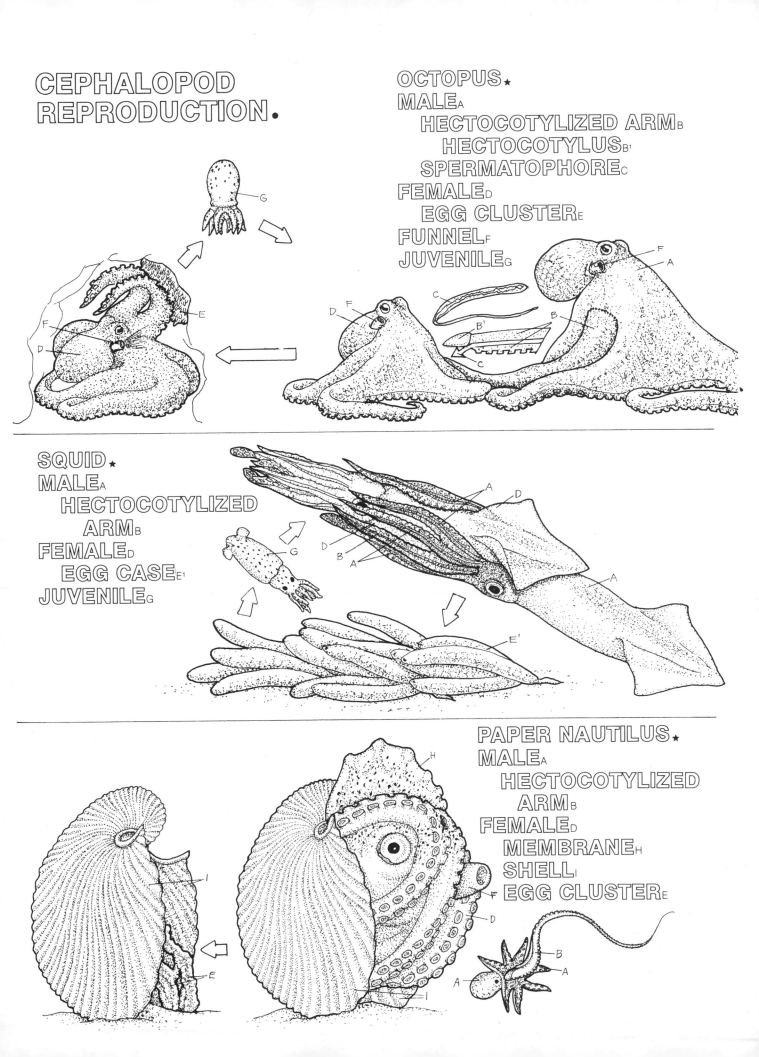

66

REPRODUCTION IN CRUSTACEANS: BARNACLE AND COPEPOD LIFE CYCLES

Reproduction presents a problem for attached organisms: How do they get together with others of the same species to mate and produce a new generation? Some species, such as oysters (Plate 63), simply release their gametes into the water, where fertilization occurs without contact between adults. However, this method is costly in the number of gametes wasted. In this plate the reproductive methods and life cycles of barnacles and copepods are discussed.

Begin coloring the barnacle life cycle with the cross section of the adult at the top of the page. Continue with the larval forms; remember that these larvae are shown greatly exaggerated in size relative to the adults. Choose a light color for the carapace (L) of the cyprid larva, as it is transparent and the body and appendages can be seen through it.

Barnacles do not release their gametes into the sea. When sexually mature, an adult barnacle uncoils its long, flexible, tubular *penis* and probes about for a nearby mate. Since barnacles are gregarious and larvae settle near their own kind, a receptive neighbor is generally available. The barnacle inserts the *penis* and transfers the sperm to its mate, as shown at the top of the plate. Barnacles, like nudibranchs, are hermaphroditic, so any adult can serve as a mate to any other adult of the same species. Fertilized *eggs* are brooded within the shell of the adult until they develop into nauplius larvae. A single individual may brood and release as many as 13,000 larvae.

The nauplius is clearly a crustacean, with *eye* spot, *antennae,* jointed *appendages,* and a shield-shaped *body.* Like any crustacean, the barnacle nauplius molts (sheds its exoskeleton) as it grows and develops. After several molts as a nauplius, a cyprid larva emerges. The cyprid does not feed. It is equipped with large *antennae* and more *appendages* than the nauplius, and the *body* is enclosed in a hinged *carapace.* After a short time, the cyprid begins to crawl about the substratum, testing for an appropriate place to settle. Cyprids are attracted to rough surfaces and

to adult barnacles of their own species. They recognize these features of their environment with tactile (touch) and chemical (''taste/smell'') sense organs in their *antennae.* If the cyprid fails to find a suitable substratum it can swim off and search further, often delaying settlement for several days (see Plate 80).

When the cyprid does locate and select a substratum on which to settle, it attaches using special cement glands in its *antennae.* The cyprid then molts and rotates its *body* such that the *appendages* are facing upward. The *appendages,* now called *cirripeds* (see Plate 28), are long and feathery; they are used by the adult to filter food from the water and carry it to the *mouth.* The cyprid's *carapace* serves as a form around which the barnacle's plates or shells are secreted. The *antennal* cement glands serve to anchor the bottom or *basal plate* to the substratum. The *fixed plates* are attached to this *basal plate* and articulate with the *movable plates,* which swing open to allow protrusion of the *cirripeds.*

Color the copepod life cycle. Remember that the size of the larval forms is greatly exaggerated.

Copepods also pass through a nauplius larval stage. During reproduction, a male copepod grasps the female with its large *antennae* and transfers the sperm to her genital opening (not shown). Most copepods brood fertilized eggs in special pouches until the larvae are released. However, in some of the common planktonic copepods (calanoid copepods), the fertilized eggs, or *zygotes,* are released singly into the water, where they hatch as copepod nauplius larvae (I). The nauplius undergoes several molts (II, etc.) and develops into a copepodite stage. The copepodite looks somewhat like the adult, but it is smaller, has fewer *appendages,* and its abdomen is not yet clearly segmented (see Plate 28). After, usually, five copepodite stages, (Stage V is shown here) the adult form is reached. The entire cycle from egg to adult may take as little as a week or as much as a year, depending on the species involved.

BARNACLE AND COPEPOD LIFE CYCLES.

SHELL PLATES ★
  BASAL A
  FIXED B
  MOVEABLE C
TESTIS D
  PENIS E
EGGS F
ANTENNA G
BODY H
APPENDAGE I
  CIRRIPED I'
GUT J
  MOUTH J'
EYE K
CARAPACE L

BARNACLE ★

CROSS SECTIONAL VIEW

VIEW FROM ABOVE

CYPRID LARVA ★

SIDE VIEW

NAUPLIUS ★

COPEPOD ★
ANTENNA G
EYE K
BODY H
APPENDAGE I
ZYGOTE M

NAUPLIUS I ★

COPEPODITE V ★

NAUPLIUS II ★

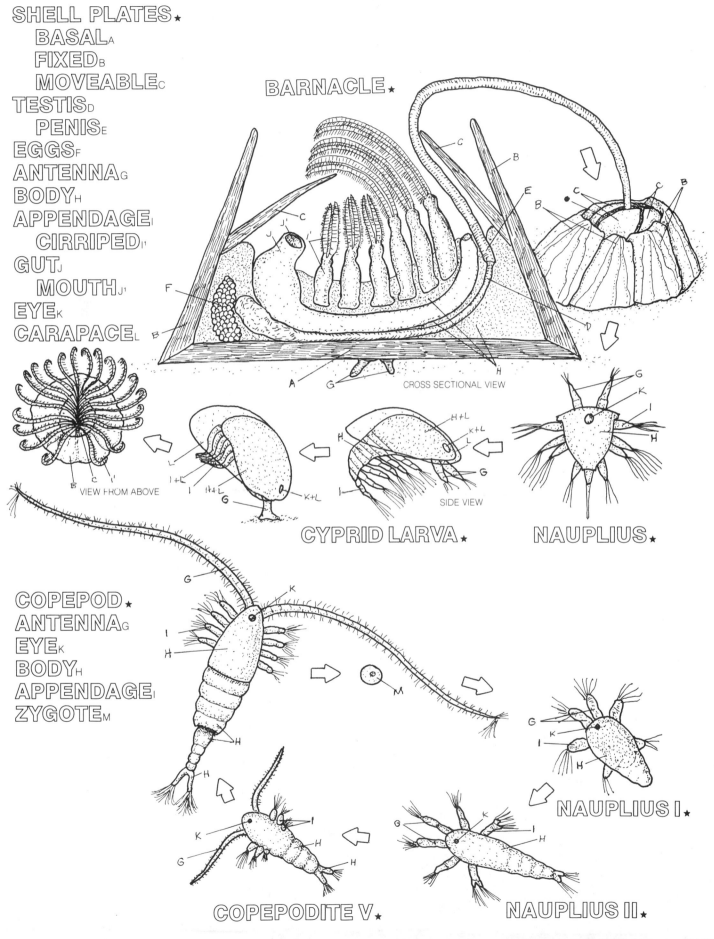

67

REPRODUCTION IN CRUSTACEANS: AMPHIPODS, STOMATOPODS, AND DECAPODS

The reproductive cycle of most crustaceans, as with barnacles and copepods (see Plate 66), involves the transfer of sperm from the male to the female followed by brooding of the zygotes for varying lengths of time. The females may carry the embryos and release them as early nauplius larvae, some later larval form, or as fully developed juveniles. Some of this variability is exemplified by the crustaceans discussed in this plate.

Begin by coloring the amphipods. Note that the female gammarid amphipod (A) has released a chemical scent called a pheromone (B), which has attracted the male. The area around the gammarid amphipods can be lightly colored to indicate the presence of the pheromone in the water. Color the rest of the crustaceans as they are mentioned in the text. Note that the larval forms are exaggerated in size relative to the adults.

Because the male must transfer sperm to the female at a time when she is sexually mature and ready to lay eggs, it is important that he be made aware of her condition. In many types of crustaceans (including the gammarid amphipod) the *female* releases a chemical (called a *pheromone*) into the water when she is approaching sexual readiness. This *pheromone* is sensed by the *male* and he is attracted to the *female*. In many species of *amphipods* the *male* holds on to the maturing *female* as shown in the illustration. The *male* releases his mate as she goes through a final molt, and then transfers sperm to her. *Female* amphipods have a *brood pouch* in which the *male's* sperm is collected; this *pouch* is easily seen in the drawing of the highly modified *caprellid* amphipod. The *female* releases her eggs into this *pouch,* where they are fertilized and retained until they hatch as juveniles (not shown). Thus, amphipods do not have planktonic larvae, and the young are released directly into the adult habitat.

Female mantis shrimps, or *stomatopods,* are also protective mothers. After the male deposits his sperm in a special pouch on the female's body (not shown) she releases her *eggs* (up to 50,000 of them!) and sticks them together in a mass about the size of a walnut. The *egg* mass is carried by the front legs and is carefully cleaned and rotated by the *female*. The *female* carries the *eggs* for several weeks and does not feed during this time. After this brooding period the *eggs* hatch as stomatopod *zoea larvae,* which act as aggressive predators feeding on other zooplankters. After a few months the *zoea* settles, metamorphoses, and takes up adult life on the bottom.

In the decapod crustaceans (crabs, shrimps, lobsters) the adults copulate, and sperm are transferred from the male to the female. Shrimp contact one another at a right angle, as illustrated. The *male* forms packets of sperm *(spermatophores)* that he attaches near the *female's* reproductive openings (not shown). In some shrimp species the female bears a special pouch in which to store the sperm for later fertilization. In others, the eggs are extruded and fertilized soon after sperm transfer. With the exception of one large group of shrimp (the penaeids), *female* decapods brood their fertilized *eggs* by attaching them to their swimmerets (abdominal appendages). Brooding *females* are referred to as being "berried females," or "in berry." After some time of brooding (duration varies from weeks to months according to species) the larvae are released as planktonic *zoea larvae*. The shrimp *zoea* illustrated here remains in the plankton for several weeks before metamorphosing to the adult form.

The larval forms of crustaceans vary considerably. The stomatopod and shrimp *zoea larvae* have recognizable features of the adult forms, but not so in the case of the spiny lobsters. Shown here is a berried *female* California *spiny lobster*. The larval stage of this animal is released after about two months of brooding by the *female*. These larvae are called *phyllosoma* (leaf body) *larvae;* they are extremely flattened, almost paper-thin, and nearly transparent. The *phyllosoma* of the California *spiny lobster* remains planktonic for up to six months before settling. Only then does it assume its adult form.

AMPHIPODS, STOMATOPODS, AND DECAPODS.

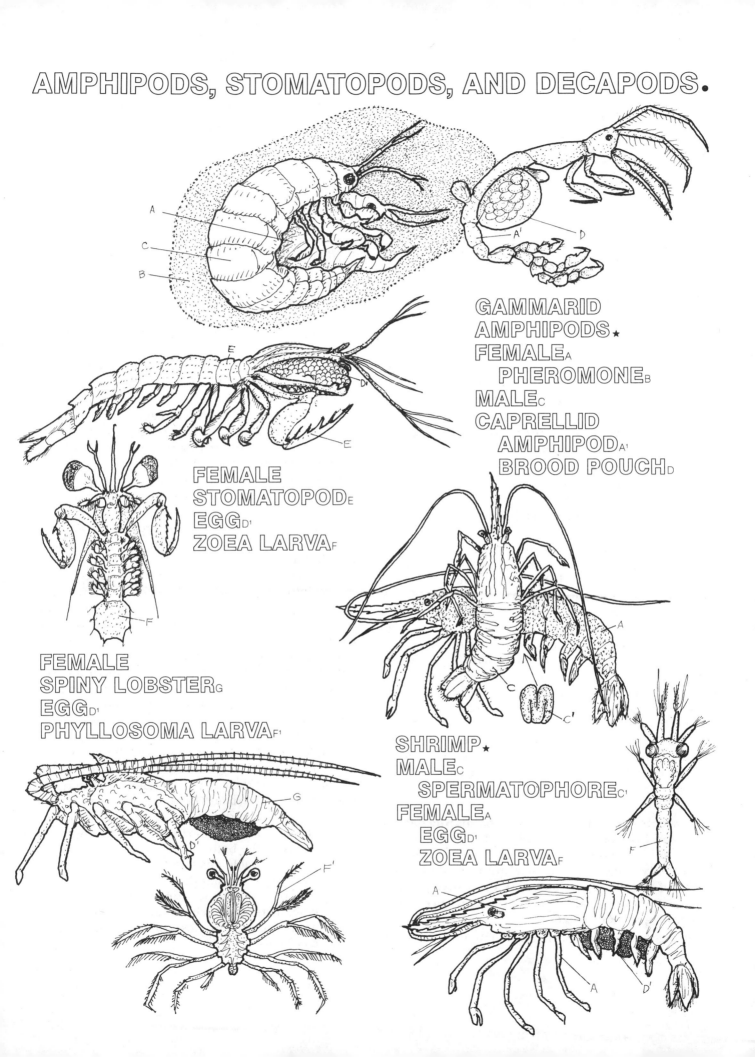

GAMMARID
AMPHIPODS ★
FEMALE A
  PHEROMONE B
MALE C
CAPRELLID
  AMPHIPOD A¹
  BROOD POUCH D

FEMALE
STOMATOPOD E
EGG D¹
ZOEA LARVA F

FEMALE
SPINY LOBSTER G
EGG D¹
PHYLLOSOMA LARVA F¹

SHRIMP ★
MALE C
  SPERMATOPHORE C¹
FEMALE A
  EGG D¹
  ZOEA LARVA F

68

REPRODUCTION IN CRUSTACEANS: CRAB LIFE CYCLE

The group of decapods, known as the "true crabs" (section Brachyura, see Plate 30), are considered advanced crustaceans, and their reproductive biology is quite complex. The life cycle of the red rock crab is discussed here as an example of reproduction in this group.

Begin by coloring the male and female in ventral (underside) view at the top of the page. Locate and color each structure and body region as mentioned in the text. Note that only a portion of the crab is colored here, and that the abdomen is shown bent away from its normal position against the thorax.

Adult *male* and *female* crabs vary in the form of their *abdomens*. Although both sexes have reduced *abdomens* (when compared with lobsters and shrimps, see Plate 29) that are normally kept flexed beneath the *thorax,* the *male's abdomen* is much narrower than that of the *female.* The abdominal appendages (modified *swimmerets*) are visible in the *male* as two pairs of shaftlike processes, the smaller pair fitting into grooves on the larger pair. At the bases of the last pair of walking legs of the *male* are two small openings (not shown) for the release of sperm from the reproductive tract. Sperm pass from the reproductive openings to the grooves on the larger pair of *swimmerets.* The smaller *swimmerets* push the sperm along the groove into the *genital openings* of the *female* as described below.

The *female genital openings* are located on the *thorax,* and the broad *abdomen* bears several pairs of feathery *swimmerets* on which the *egg cluster* is held during brooding.

Now color the illustrations of the mating behavior and life cycle of the crab. Color each stage as it is mentioned in the text. The natural color of the early egg cluster (I) is salmon/coral pinkish, and of the late egg cluster (J), brownish purple. Note that the larval and juvenile stages are greatly exaggerated in size.

During the life cycle of the rock crab, the crabs must molt (shed their exoskeleton) in order to grow. The crab accomplishes this by resorbing much of the calcium salts from its old skeleton and then swelling with water to burst the old skeleton. The new skeleton is soft, allowing the crab to crawl free from the old encasement. Once free, the crab continues to swell and increase in size, then gradually hardens the new skeleton by the redeposition of the conserved salts. Typically, the *male* molts before the mating season. The *female,* however, must be newly molted for successful mating to occur. It is critical that the *male* discover the *female* prior to her actual molting so that he can be present to fertilize her eggs before her new exoskeleton hardens. The *female* probably releases a chemical attractant in her urine at the appropriate time. Once together, the *male* grasps the *female* and holds her until she begins her molt. He releases her as she emerges from her old exoskeleton or *"molt"* (top right diagram of the life cycle). After her new exoskeleton becomes sufficiently firm, she lowers her *abdomen* and allows the *male* to insert his modified *swimmerets* into her *genital openings* for sperm transfer. Once this has been accomplished the *male* holds the *female* in a protective embrace until her exoskeleton hardens.

The eggs are fertilized by the *male's* sperm as they are released from the *female's genital openings.* The sticky fertilized eggs adhere to one another, and the *female* attaches them to her "hairy" *swimmerets.* A *female* rock crab may carry several thousand eggs in her *egg cluster.* The mass of eggs is light in color when newly formed, but darkens substantially as the embryos develop. After several weeks the young crab larvae break free of their egg membranes and emerge as planktonic *zoea larvae.* The *zoea larvae* swim, feed on zooplankton, and molt several times until they reach a crablike *megalops larval* stage. The *megalops* finally sinks to the bottom, and at its next molt becomes a recognizable *newly settled rock crab. Juveniles* molt frequently and the species discussed here may often be seen in the striped phase, as shown. These crabs mature in three to four years and are then ready to participate in this complex adult reproductive behavior.

CRAB LIFE CYCLE.

MALE A
MOLT B

ABDOMEN C
THORAX D
SWIMMERETS E

FEMALE F
MOLT G
GENITAL OPENING H

EGG CLUSTER ★
EARLY I
LATE J
ZOEA LARVA K
MEGALOPS LARVA L
NEWLY SETTLED
CRAB M
JUVENILE N

69

REPRODUCTION IN ECHINODERMS: ECHINODERM LIFE CYCLES

The echinoderms are almost all bottom-dwelling, slow-moving animals, and usually depend upon planktonic larval forms for dispersal. However, there are some notable exceptions to this scheme. This plate discusses the typical life cycle of a sea urchin, and the rather unusual life cycle of a small sea star.

Color the sea urchin life cycle, coloring each stage as it is mentioned in the text.

Sea urchins usually occur in large aggregations (see Plate 3). This gregarious habit allows urchins to "broadcast" their *sperm* and *eggs* into the water, in turn stimulating the spawning of other individuals.

After fertilization, the *zygote* is free in the plankton, and develops in a few days into an *early echinopluteus larva*. This *larva* has long, ciliated *arms* for locomotion and the collection of phytoplankton, on which it feeds. This stage may persist for days or months, depending on the species of urchin, but eventually begins to change form (metamorphose). The *late echinopluteus larva* has reduced *arms* that are gradually resorbed by the body as the *juvenile* urchin develops. When the *juvenile* settles out of the plankton, it is only about one millimeter across, with just a few spines and tube feet. It grows quickly, however, and soon looks like a miniature adult.

Color the typical sea star larval forms in the boxed area. These represent the planktonic larval forms of non-brooding sea stars (not shown).

Most sea stars undergo a life cycle that is similar to that of the sea urchin. The zygote (not shown) develops into a *bipinnaria larva* possessing winglike folds that are ciliated for swimming. This stage changes into a more elaborate *brachiolaria* larva, with elongate *arms* and three *pre-oral arms* used for attachment when the *brachiolaria* settles and metamorphoses to a young sea star.

Now color the modified life cycle of the six-rayed sea star as it is discussed in the text. Note that the brooding female is shown with her body covering the juveniles (E¹), which receive the colors of both the mother (B²) and the juvenile (E¹).

In this sea star the *female* retains and cares for her *eggs* while they develop. She attaches to the bottom or side of a rock by the tips of her arms and forms a cup-shaped brooding area with her oral surface. The *female* lays her *eggs* into this pocket, and the *eggs* are fertilized by *sperm* from nearby *males*. After fertilization the *female* begins a two-month vigil during which she remains in this brooding position, moving only to manipulate the developing *zygotes* to keep them clean and well aerated. The brooded embryos emerge from their egg membranes as drastically modified *brachiolaria* larvae (the *bipinnaria* stage is bypassed completely). This *brachiolaria* lacks the long *larval* arms seen in the more typical *planktonic* form, and bears only the three adhesive *pre-oral arms*. The *larval body* begins to change shape, and soon five distinct bulges appear. These bulges develop into five of the sea star's arms, and a sixth soon forms. The *juvenile* sea star develops tube feet and the *larval body* is resorbed. After two months of confinement, while developing in the brooding area, the *juveniles* begin to move about. The *brooding mother* re-attaches her entire oral surface to the substratum, but remains in place with her young for a few more days. Soon the small (1 mm) *juveniles* become more active and leave the "nest" forever. This type of life cycle insures a steady input of young sea stars into the adult habitat, but sacrifices the benefits of a planktonic larval stage (see Plate 60).

ECHINODERM LIFE CYCLES.

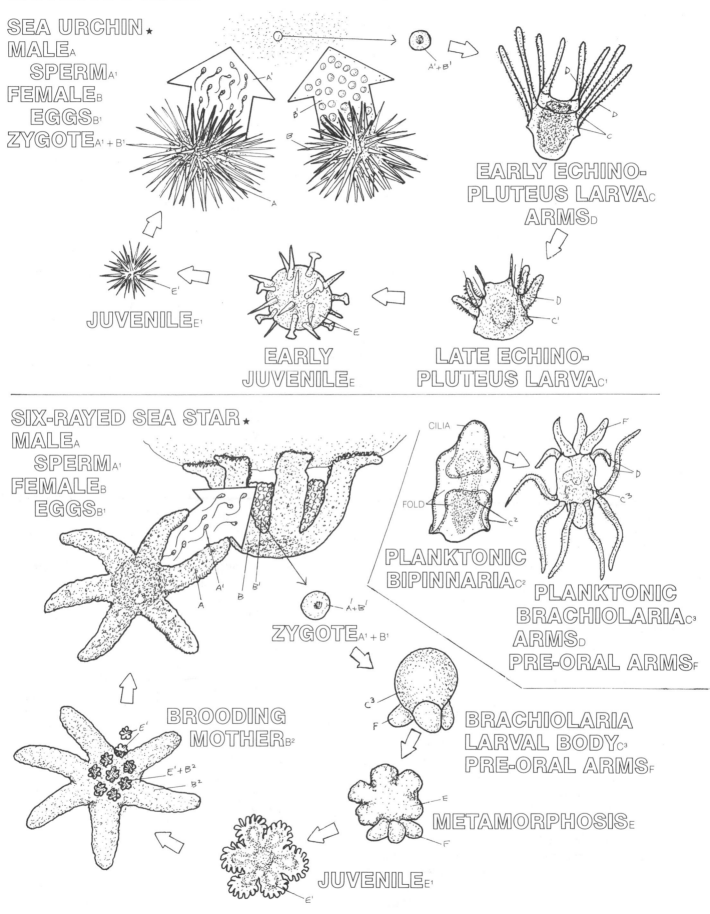

SEA URCHIN ★
MALE A
  SPERM A'
FEMALE B
  EGGS B'
ZYGOTE A' + B'

EARLY ECHINO-
PLUTEUS LARVA C
ARMS D

LATE ECHINO-
PLUTEUS LARVA C'

EARLY
JUVENILE E

JUVENILE E'

SIX-RAYED SEA STAR ★
MALE A
  SPERM A'
FEMALE B
  EGGS B'

ZYGOTE A' + B'

CILIA

FOLD

PLANKTONIC
BIPINNARIA C²

PLANKTONIC
BRACHIOLARIA C³
ARMS D
PRE-ORAL ARMS F

BRACHIOLARIA
LARVAL BODY C³
PRE-ORAL ARMS F

BROODING
MOTHER B²

METAMORPHOSIS E

JUVENILE E'

70

REPRODUCTION IN ELASMOBRANCHS: SHARKS AND SKATES

Begin by coloring the shark's pelvic fin (A) and clasper (B); use the male color for the body. Next color the copulating (mating) sharks using different colors for the male and female. Note that the adult skate and the horn shark are females and receive the female color (D).

Sharks, skates, and rays practice internal fertilization, which is a more efficient and less wasteful means of fertilization than releasing gametes into the environment. To accomplish the transfer of sperm, male elasmobranch fishes have special intromittent organs (transfer organs) called *claspers.* The upper drawing shows the *clasper* of a horn shark; the *clasper* is derived from the shark's *pelvic fin.* During copulation, the *clasper* is inserted into the female's genital opening and the *spur* is erected to insure that the sharks stay coupled long enough for sperm transfer to occur. In the spotted dogfish sharks, the *male* wraps his body around the *female,* inserts a clasper and transfers his sperm. In other sharks that are less supple than the spotted dogfish, the male and female lie side by side (not shown). The male holds onto the female's pectoral fin, erects the clasper at a right angle to his body, and inserts it into the female.

Color the skate and its egg case, embryo, and the yolk sac. Note that in the drawing on the right the embryo is large and well developed. Note also that only the intact egg case is to be colored. Color the horn shark and its egg case.

Once the eggs have been fertilized within the female, one of three possible patterns of development occurs, depending on the species of fish. In one pattern, the fertilized eggs are packaged in a special egg case and released by the female to develop outside her body. In this situation each egg is provided with a large amount of yolk to nourish the developing *embryo* during its growth. Skates employ this reproductive strategy that frees the female from prolonged maternal care. The illustration at the lower left of the plate shows a female *skate* swimming away from her large (23 cm) *egg case,* sometimes called a "mermaid's purse." Inside this *egg case* are one or several developing *embryos.* The two drawings of opened *egg cases* show the *embryos* increasing in size and their *yolk sacs* dwindling as the

yolk is used for nourishment. Under normal conditions the *egg case* gradually deteriorates and begins to fall apart as the young skate is ready to emerge and begin a life of its own.

Some sharks also produce an *egg case,* exemplified by the spiral *egg case* of the *horn shark* shown here. The young sharks emerge from this *case* when they are about 12 centimeters long and appear as miniature adults.

A second developmental pattern seen in elasmobranchs is the retention of developing fertilized eggs inside the female's reproductive tract (not shown). Upon completing their development the young are released as fully formed juveniles. This type of development is known as ovoviviparity (ovo: egg, viviparity: live birth). These eggs are still provided with a yolk sac that serves as their source of nourishment, the female simply protecting the embryos within her body.

Now color the embryonic shark in its placenta-like arrangement within the mother's uterus. Note that the "placenta" and umbilical cord receive the same color as the yolk sac (G) from which they are derived.

Some sharks undergo a third type of development known as viviparity. In this situation the mother provides the developing *embryos* with nourishment in addition to the yolk within the egg. Supplying this nourishment may occur in a number of different ways. The drawing shows a method that resembles the mammalian placental connection between female and embryo. A smoothhound shark *embryo* is shown folded in an *embryo sac* within its mother's *uterus.* An *"umbilical cord"* and *"placenta"* containing embryonic blood vessels connect the *embryo* to the wall of the *uterus.* The *"placenta"* receives nourishment from the mother's circulatory system, and it is transferred to the developing *embryo* through the *"umbilical cord."* Actually, the *placenta* here is a modified *yolk sac,* and the *"umbilical cord"* is the elongated connection between the *yolk sac* and the *embryo.* The result of this arrangement is the direct provision of nutrients by the mother to the *embryo.*

SHARKS AND SKATES.

PELVIC FIN_A
CLASPER_B
SPUR_{B¹}

COPULATION ★
MALE_C
FEMALE_D

SKATE_{D¹}
EGG CASE_E
EMBRYO_F
YOLK SAC_G

HORN SHARK_{D²}
EGG CASE_{E¹}

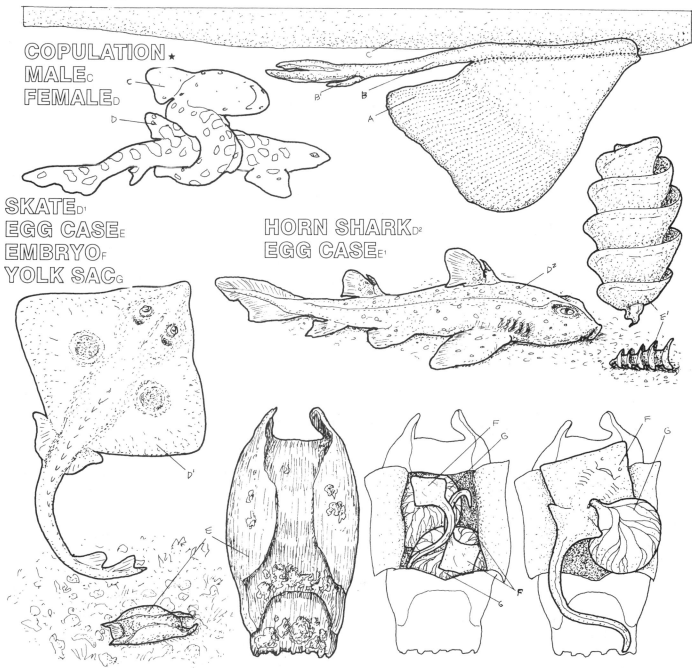

UTERUS_H
EMBRYO SAC_I
EMBRYO_{F¹}
PLACENTA_{G¹}
UMBILICAL CORD_{G²}

"PLACENTAL" ARRANGEMENT ★

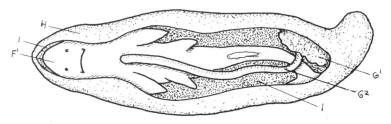

71

REPRODUCTION IN BONY FISHES: LIVE BEARERS AND BROODERS

Bony fishes employ a variety of reproductive strategies to insure continuation of their species. Some of these methods of reproduction are discussed in this and the next two plates.

Parental care of the young is expensive in terms of the parents' time and energy, but it does tend to insure the survival of a high percentage of offspring. The strategy is to produce relatively few young—that is, a number that can be cared for—and protect them until they are able to take care of themselves (a very different situation from the free spawning of various invertebrates, as discussed earlier). This approach also insures that juveniles will be "turned loose" in the proper habitat.

Color the various fishes gray except for the areas located within circles and the seahorse's pouch. These special areas receive one of four different colors (A–D). Within the large circles below each fish, color only those structures outlined with a heavy line. Do not color the lightly drawn material. Begin with the live-bearing surfperch and then color each fish separately as it is treated in the text.

As seen in the previous plate, all of the elasmobranch fishes (sharks, rays, skates) practice internal fertilization followed by some protection of the developing young, either within the female or in an elaborate egg case. The retention of developing young within the parent's body (or in some special brooding area on the parent) is the most effective way to insure the safe development of the offspring. As mentioned above, however, it is costly to the adults engaged in such activities. This type of reproduction is practiced by several families of bony fishes, most of which are called *live bearers.*

The surfperches, common along the Pacific coast of North America, are classic examples of *live bearers.* The female surfperch is inseminated by the male following rather complicated mating behaviors. The fertilized eggs are retained in the female's ovary where they develop to miniature adults. During development, nourishment is provided by the mother in the form of a nutrient-rich secretion that is absorbed by the young. The fins of the embryos are enlarged and richly supplied with blood vessels (vascularized) for this function.

A number of other strategies involve protecting the developing embryos either in some sort of nest (see next plate) or by means of a special brooding area on the adult fish. The gafftopsail catfish of the Gulf of Mexico is an example of a *mouth brooder.* As the name indicates, the fertilized eggs are carried (brooded) within the parent's mouth, in this instance the male's. The male carries from 40 to 60 marble-sized eggs in his mouth for about nine weeks, while the young develop to the hatching stage. After hatching, the juveniles may stay with the father for an additional month, during which time they enter and leave the mouth at will, thus remaining protected until finally venturing off on their own. This exceptional gesture of parenthood is especially impressive in that the male catfish does not eat during the entire brooding period.

There is a variety of other brooding methods among the bony fishes. The *Kurtus* of the South Pacific uses its head. Again it is the males which carry the embryos; they are called *forehead brooders.* The male *Kurtus* has a special hook on its forehead to which he attaches a mass of fertilized eggs. In this case the adult can continue to feed while he protects the embryos until they hatch.

A very effective brooding technique is seen in the seahorses and pipefishes. In these small, slender fishes, the female lays her eggs into a special *pouch* on the ventral surface of the male, who then fertilizes and broods the eggs, out of harm's way. The young remain within this *pouch,* nourished by their father's blood supply, until they are fully formed young juveniles. The male then "gives birth" by writhing to and fro and flexing the *pouch* muscles, forcing the young fish out on their own.

LIVE BEARERS & BROODERS.

LIVE BEARER:A
MOUTH BROODER:B
FOREHEAD BROODER:C
POUCH BROODER:D

SURF-
PERCH ★

GAFFTOPSAIL
CATFISH ★

KURTUS ★

SEA-
HORSE ★

Many species of bony fish prepare some type of nest or protected area in which to deposit their fertilized eggs. Three very different types of nesting behavior are discussed in this plate.

Reserve bright blue for the male (I) and dull blue for the female (J) damselfish. Begin by coloring the nesting female and spawning male salmon using bright red (A) for the bodies and grayish green (B) for the heads. Do not color the jaw, pelvic fins, pectoral fins, ventral (bottom) surface, or the edge of the caudal fin in the adult salmon. Color the other stages in the life cycle.

Nest building by the sockeye salmon is only a small part of this fish's incredible life cycle. After spending from two to four years at sea in the North Pacific, the adult sockeye return *to* their home *stream* in the late summer to reproduce (spawn). This journey involves a migration from the ocean, through estuarine areas, into freshwater rivers and streams. Some of these fish travel up to 1,500 miles to eventually return to the small tributaries where they themselves were hatched years earlier. Exactly how the salmon accomplishes this feat is still not fully understood. It appears that the fish use several different sensory cues during their migration.

As the adult sockeye salmon move upstream, their silver-blue sea-run colors change to bright red mating colors. The *male's* jaws become grotesquely hooked, and he develops a pronounced hump on his back just ahead of the dorsal fin. When the *female* reaches her home stream, she uses her tail to prepare a shallow trough or nest in the gravel bottom. The *male* joins her over the nest, and they simultaneously release sperm and *eggs,* which drop into the depression. After a short time, the *female* moves upstream and digs another nest, the materials from which are carried downstream by the current and cover the *eggs* retained in the first nest. The female may prepare a half-dozen or more such nests until she has released her 3,000 or more eggs. The combined series of nests with their fertilized eggs is called a "redd." After migrating and spawning, the formerly sleek sockeyes are emaciated and very weak; their mission completed, they soon die.

The fertilized salmon *eggs* remain buried under several inches of sand and gravel through the winter. By spring each embryo has developed to a 2.5 centimeter-long *alevin* that still carries the remainder of the egg's yolk in the attached *yolk sac.* The *alevin*

grows into a *parr,* which remains in fresh water for about two years. Toward the end of this period it matures to a 15 centimeter *smolt,* which moves downstream and out to sea where it feeds and grows to adulthood.

Color the grunion spawning cycle. These figures represent a bird's eye view of the edge of a sandy beach. Color in the stippled area (H), the fish, and the circles representing clusters of buried eggs.

The California *grunion* lay their *eggs* high up on sandy beaches during the spring and summer and let the warm southern California sun incubate the developing embryos. For this strategy to be successful, the activity must coincide with the season's highest (spring) tides. For three or four nights following the highest tide (the highest tides are always at night during the spring and summer in southern California) *male* and *female grunion* ride inshore on *waves* at the peak of the *high tide.* As the wave recedes, the *female* wriggles tail first into the sand and one or more *males* wrap themselves around her. She releases her *eggs* into a tunnel-like "nest" she had made, and the *males* release their sperm (spawning). The adults swim off at the next *high wave* and the nest is buried by the waves' action. It is important that the *grunion* accomplish their egg laying immediately following the highest tide of a cycle so that no waves will reach the nests for at least 10 to 12 days. When the *waves* do finally wash over the nests several days later, the *juveniles* burst out of their protective membranes, wriggle to the sand's surface, and swim to sea with the receding *wave.* They will return to breed the next year as 12.5–15 centimeter adults.

Color the damselfish cycle.

As the season for reproduction approaches, the *male damselfish* takes on bright breeding colors and busies himself preparing a patch of *red algae* on the coral reef. He then attracts a prospective mate by special movements of his white-edged caudal fin. The *female damselfish* is enticed into his nest of *red algae* and releases her *eggs,* which the *male* fertilizes by releasing sperm over them. The *female* then leaves the nesting site (the *male* may even chase her away), and the *male* begins a several-week-long vigil, guarding the *eggs* and fanning them with his fins to keep them clean and well oxygenated during development.

NEST BUILDERS.

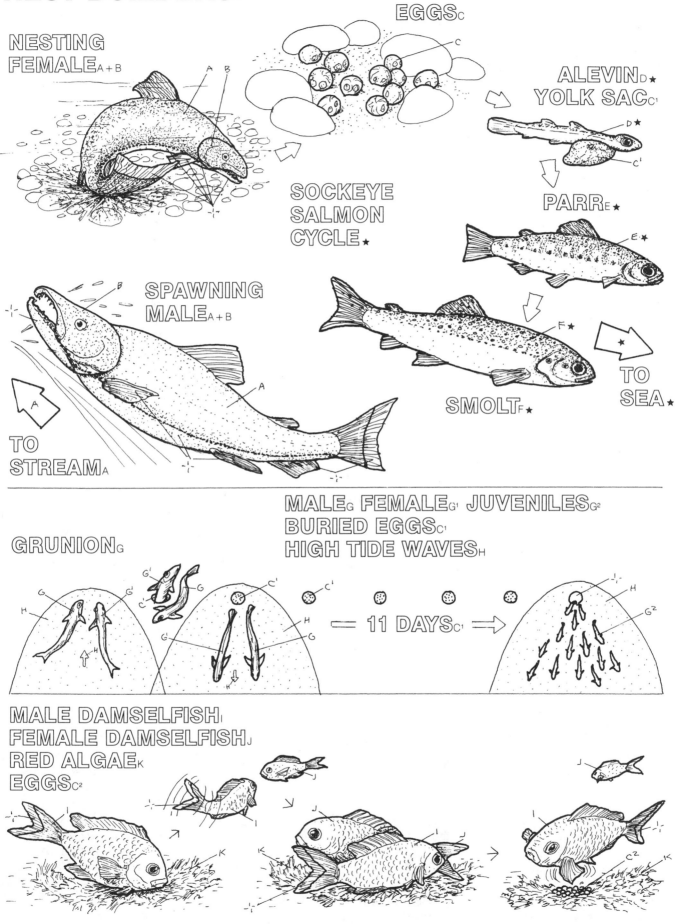

NESTING
FEMALE$_{A+B}$

EGGS$_C$

ALEVIN$_D$★
YOLK SAC$_{C^1}$

SOCKEYE
SALMON
CYCLE ★

PARR$_E$★

SPAWNING
MALE$_{A+B}$

SMOLT$_F$★

TO
SEA ★

TO
STREAM$_A$

MALE$_G$ FEMALE$_{G^1}$ JUVENILES$_{G^2}$
BURIED EGGS$_{C^1}$
HIGH TIDE WAVES$_H$

GRUNION$_G$

11 DAYS$_{C^1}$

MALE DAMSELFISH$_I$
FEMALE DAMSELFISH$_J$
RED ALGAE$_K$
EGGS$_{C^2}$

73

REPRODUCTION IN BONY FISHES: BROADCASTERS

Many marine fishes neither carry their young nor build nests for their protection. Instead, these fishes release their gametes (sperm and eggs) into the water, where fertilization and development take place without further parental involvement. Such fishes are commonly called "broadcasters," as their gametes are distributed, or broadcast, by the waves and currents.

A number of factors contribute to the success or necessity of broadcasting as a reproductive strategy. Pelagic fishes (see Plate 36) are generally not associated with a suitable substratum for nest building and cannot stop swimming to brood their young. The release of gametes directly into the water takes advantage of the ocean's free dispersal service and the rich supply of plankton potentially available to young fishes (see Plate 60). The reproductive habits of two species of broadcasting fishes are discussed in this plate.

Begin by coloring the male and female herring and their sperm and eggs. The circled enlargement is a portion of the algal substratum upon which fertilized eggs have been deposited. The larval herring are not drawn to scale with the adults.

The northern herring was discussed in Plate 36. This fish is found in the Atlantic Ocean, and a closely related subspecies occurs in the Pacific. Herring range widely in both oceans, often in shoals of millions of fish. The herring is a pelagic fish that feeds on zooplankton and usually lives over deep water. It is a broadcaster and usually moves into shallow water to spawn. Along the Pacific coast of the United States herring come into estuaries and shallow bays. *Females* release their *eggs* near males, which then release *sperm*. The fertilized eggs are heavier than water (termed demersal eggs) and sink to the bottom, where they adhere to *algae* and eel grass. Within 10 to 15 days the *attached embryos* hatch as *yolk-sac larvae* at a length of 6 to 8 millimeters. The *yolk-sac larvae* remain near their hatching site for a few days, feeding on reserves of yolk. As they grow into larger *larvae* (29 mm), they swim upward into the water, where currents carry them away. The herring *larvae* continue to grow into *prejuveniles* (about 40 mm) that soon join other herring in the open water. These young fish will mature to spawn in 2 to 7 years.

Now color the diagram of the migration routes and life history stages of the European eel. The

small eel pictured on the map receives the separate colors of the three advanced stages of the eel: glass (K), yellow (M), and silver (N), which occur in and around the coastal streams of Europe. The map of the Atlantic Ocean includes the coastal waters and streams inhabited by the American eel; this darkly stippled band and the arrows indicating migration routes receive the color of the title: American eel (O).

The common or European eel is a catadromous fish; that is, it lives in fresh water and migrates to the ocean to spawn (the reverse of salmon, which are termed anadromous; see Plate 72). The migration of these eels is truly amazing. In European fresh water *streams* the adults change from the common *"yellow eel"* to the silver-white *"silver eel,"* with enlarged eyes and reduced mouths. These *silver eels* migrate across the Atlantic Ocean to an area known as the *Sargasso Sea,* a floating "island" of the seaweed *Sargassum* (a brown alga), located in a calm area of the western Atlantic near Burmuda. Details of the journey are not known, but once at their destination the *silver eels* spawn in deep water (500 m) and die. The *eggs* slowly float to the water's surface, and an exotic-looking larva called the *leptocephalus* (thin head) emerges in the spring. This transparent, leaf-shaped *larva* then sets off back toward Europe, a migration of more than 2,500 miles that takes two years or more to complete. As it swims and drifts eastward in the Gulf Stream current it grows much larger, becoming a 7.5 centimeter *"glass eel,"* or elver, by the time it reaches the European coast. The *glass eel* enters *coastal streams* and acquires the yellow adult pigmentation. Young eels migrate upstream by the thousands, sometimes over a distance of many miles. The urge to migrate comes again when the eels are anywhere from four to twenty years old, and the adults begin their journey back to the *Sargasso Sea* to breed.

The life cycle of the European eel is still imperfectly understood. Adult eels have never been caught offshore in the Atlantic along the proposed migration route. The entire migration has been inferred from the distribution of the *leptocephalus larvae*; small ones are found in the *Sargasso Sea,* and progressively larger ones are found as one moves eastward towards Europe. There is some question whether the European eel and *American eel* are separate species. *American eels* spawn in the same area of the *Sargasso Sea,* but their larvae reach the coast of North America in only one year.

BROADCASTERS: HERRING.

MALE_A SPERM_{A1}
FEMALE_B EGG_{B1}
ALGAL SUBSTRATE_C

ATTACHED EMBRYO_D
YOLK-SAC LARVA_E
LARVA_F
PREJUVENILE_G

EUROPEAN EEL.

SARGASSO SEA_H
EGGS_I
LEPTOCEPHALUS LARVA_J
GLASS EEL_K

COASTAL STREAMS_L ★
YELLOW EEL_M
SILVER EEL_N
AMERICAN EEL_O

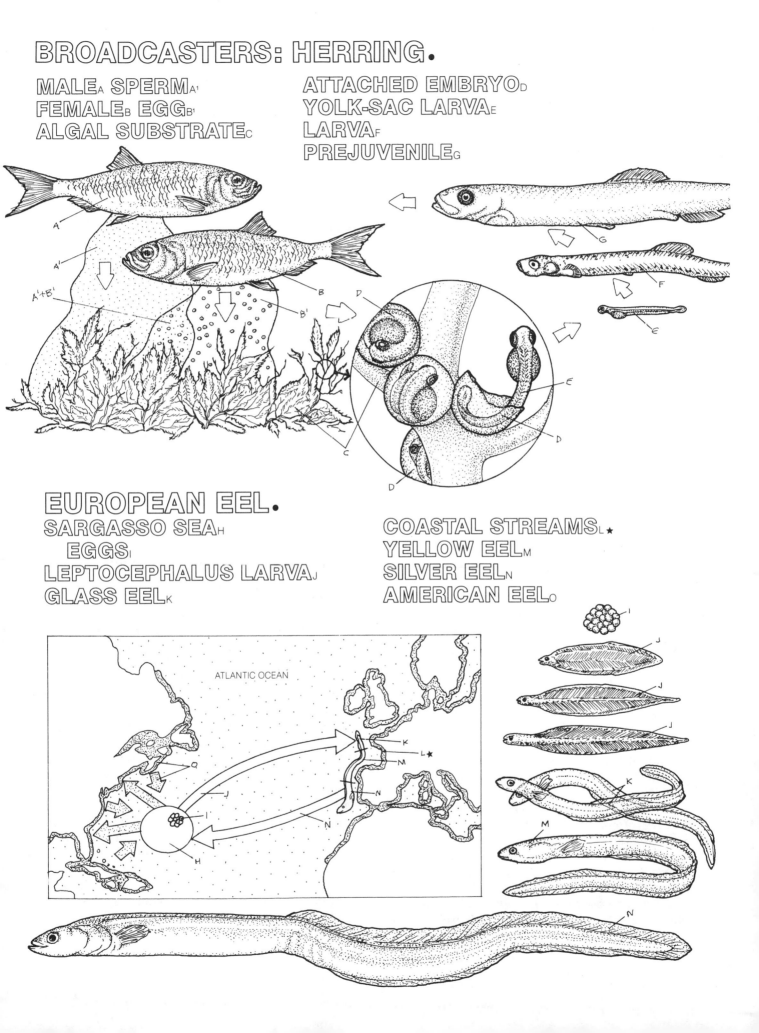

74

ANNUAL CALIFORNIA GRAY WHALE MIGRATION.

Many species of whales undertake annual migrations. These animals require tremendous amounts of food to fuel their warm-blooded metabolisms and huge bodies. Thus they travel to feed in rich polar *feeding grounds* in the spring and summer, and retreat to warmer waters during the winter to breed. The timing of the reproductive cycle of the whales is intimately associated with this annual pattern. This plate examines the annual migration of the California gray whale. This migration is well known because the gray whales often travel close to shore and spend the winter in shallow coastal lagoons where they are easily observed. Most other whale species migrate on the high seas and are far more difficult to study.

Begin by coloring the migration route of the gray whale, then color the feeding grounds and the mating and nursing sites. Then color the illustrations of the various examples of gray whale behavior as each is mentioned in the text.

In late spring the gray whales arrive on their northern *feeding grounds* in the Bering Sea and Arctic Ocean, north to the edge of the polar pack ice. Gray whales feed on benthic invertebrates (mostly small crustaceans called amphipods; see Plates 28 and 48), and are limited to rather shallow water for feeding. During the long days of the Arctic summer the gray whales feed almost continually, consuming up to a ton of food each day as they grow and replenish their blubber supplies. During this time the females wean their calves, which will make the southward fall migration on their own. In September, as the pack ice spreads over the feeding grounds, the whales begin their four-thousand-mile trek southward (the longest known migration of any mammal). The route follows the west coast of North America, and in December and January whales are frequently sighted along Oregon and California, often very close to shore. They are sometimes observed holding their heads out of water, cocked at an angle, in a behavior called *"spy-hopping."* It is speculated that they may be looking for, literally, landmarks on shore. Migrating gray whales are also observed to *breach,* leaping completely out of the water. The reason for this behavior is also unclear; it may involve visual searching by the

whales, or serve to dislodge parasites from the skin, or it might be termed play behavior.

Migrating whales swim steadily all day and perhaps through the night. After three months, they reach their destination in one of several warm-water lagoons on the western shore of Baja California, Mexico. In the relatively calm lagoons the pregnant female gives birth to a single *calf* (in rare instances to twins). A gray whale *calf* may be 5 meters long and weigh nearly a ton at birth. For the next two months the *mother* suckles her *calf* on her rich (50 percent fat content) milk. The teats of the mammary glands are located in slits in the female's underside. They are equipped with special muscles to actually force the milk into the *calf's* mouth. While nursing, the female generally lies on her side, with one fin out of the water. She is often assisted in the nursing process by another adult female called an *"aunt."* An *aunt* is a female between pregnancies (adult females give birth every two years).

Females begin to ovulate (produce fertilizable eggs) in November and may mate during the fall migration or during the winter stay in the lagoons. Mating sometimes involves a prolonged courtship during which two or more *males* pursue a *female.* When she becomes sexually receptive, she turns on her back and a *male* lies alongside, also on his back. The *male's* intromittent organ is positioned near the *female's* genital opening (not shown), and they roll together to complete the coupling. The second *male* may stand by and assist the mating couple in maintaining their position in the water. The threesome may remain together for a day or more after mating. Pregnancy in gray whales lasts thirteen months (gestation period); nursing and weaning occupy another six to eight months.

By early spring the new *calves* have grown to over 6–8 meters in length, gaining up to 100 kilograms each day, and are ready to begin the migration to the Arctic *feeding grounds* with their *mothers.* In three months the gray whales reach the *feeding grounds* where the females with new *calves* replenish their taxed energy reserves, and the pregnant females feed to nourish their unborn young.

ANNUAL CALIFORNIA GRAY WHALE MIGRATION.

MIGRATION ROUTES A
FEEDING GROUNDS B
MATING/NURSING SITES C

FEEDING B¹
SPYHOPPING D
BREACHING E
NURSING MOTHER C¹
NURSING CALF C²
NURSING AUNT C³
MATING MALE F/FEMALE C⁴
MIGRATING MOTHER A¹
CALF A²

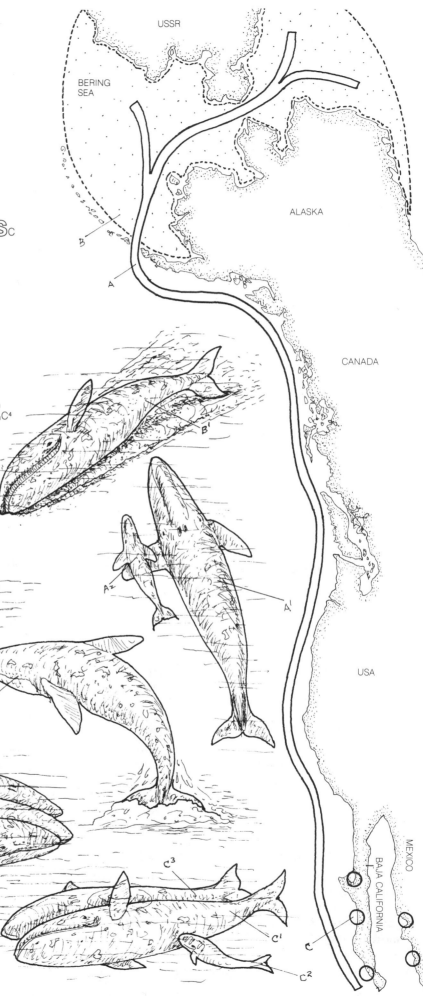

75
ELEPHANT SEAL ROOKERY

Adult elephant seals spend much of the year alone at sea. During this time the population is widely dispersed as individuals search for food. However, every year the adults congregate for breeding in certain areas called rookeries. As with all pinnipeds (seals, sea lions, and walruses, see Plate 46), elephant seal pups are born on land. The rookeries are usually located on remote islands, safe from mainland intruders and predators. An elephant seal rookery typically occupies a sandy beach, and at the peak of the winter breeding season it is crowded with seals. The breeding behavior of these animals is quite complex and apparently acts to insure that the strongest possible offspring are produced each year.

Begin by coloring the top panel, which illustrates the struggle for social position among the bull elephant seals. The most aggressive bull is designated a beachmaster (B). Next, color the second panel showing the cows beginning to arrive at the rookery. Note that the beachmaster is in the center of the rookery.

The males or *bulls* arrive at the rookery in early winter, several weeks before the arrival of the females. During this waiting period, the *bulls* fight repeatedly to establish a pecking order (hierarchy) among themselves. The strongest *bulls* remain on the beach as *beachmasters* and the rest are chased out to sea. The number of *beachmasters* depends on the overall size of the rookery. When the adult females, or *cows,* begin to arrive, the dominant *beachmaster* takes up a position in the portion of the rookery most favored by the *cows*. The lower-ranking *beachmasters* partition the remaining rookery area among themselves. The *cows* do not arrive at the rookery all at once, but over a two-to-three-month period.

Cows give birth to a single *pup* soon after arriving at the rookery (usually beginning in January). The *pups* are nursed for about four weeks and grow to a weight of 140 or 180 kilograms. The *cows* rarely feed during this nursing period and tend to remain close to their *pups*. If a *cow* is separated from her *pup,* she is able to locate and distinguish it from others by its unique smell and cry. After the *pups* are weaned, they congregate at the edges of the rookery, away from the mating adults. After weaning her *pup,* the *cow* enters a short period of sexual receptiveness (called estrus) when she will allow a *beachmaster* to mate with her. After mating, the *cows* leave the rookery and return to

the sea; little is known about their activities during the non-breeding season.

Now color the third panel from the top of the plate. This illustration depicts the rookery at the peak of the breeding season. Note that the beachmaster is confronting a challenging bull, while another bull waits at the edge of the rookery.

During a two or three month period as *cows* are arriving at the rookery, the males do not feed or return to the water. At the peak of breeding activity new *cows* are arriving to give birth while others nurse, or come into estrus and mate. The *cows* are highly gregarious, and their various activities appear almost as ripples in a sea of plump seal bodies. Adding to this confusion are constant challenges by *bulls* attempting to "dethrone" the *beachmasters*. Other *bulls* sneak into the rookery area and try to mate with receptive *cows*. The *beachmaster* must remain aware of *cows* in estrus and forcefully drive away intruding *bulls* to maintain his dominant position. Many nursing *pups* are crushed in this chaos. The cost to the *beachmasters* is also high, as they lose a great deal of weight during the breeding season. Studies indicate that *bull* elephant seals only survive two or three seasons as a *beachmaster,* and tend to die much younger than non-dominant *bulls*. They do, however, leave their marks on the new generation, since three or four *beachmasters* account for 90 percent of the mating that occurs each season. Thus, the strongest, most aggressive *bulls* father the majority of each new generation.

Now color the lowest panel showing the rookery area being utilized as a haul-out during the non-breeding season. Note the close association tolerated between mature bulls (A) and immature males (F).

When the breeding season ends in early spring, the *bulls* leave the rookery. Only the weaned *pups* remain for awhile, before going to sea to feed. During the non-breeding season the rookery is used as a haul-out or resting area primarily by *bulls* when they are not feeding at sea. During this period *bulls* tolerate or ignore one another, and it is not unusual to see a "pile" of elephant seals hauled out on the beach. As winter approaches the *bulls* return to the rookery and resume the struggle for positions of dominance.

ELEPHANT SEAL ROOKERY.

BULL_A
BEACHMASTER_B
COW_C
PUP_D
IMMATURE MALE_F

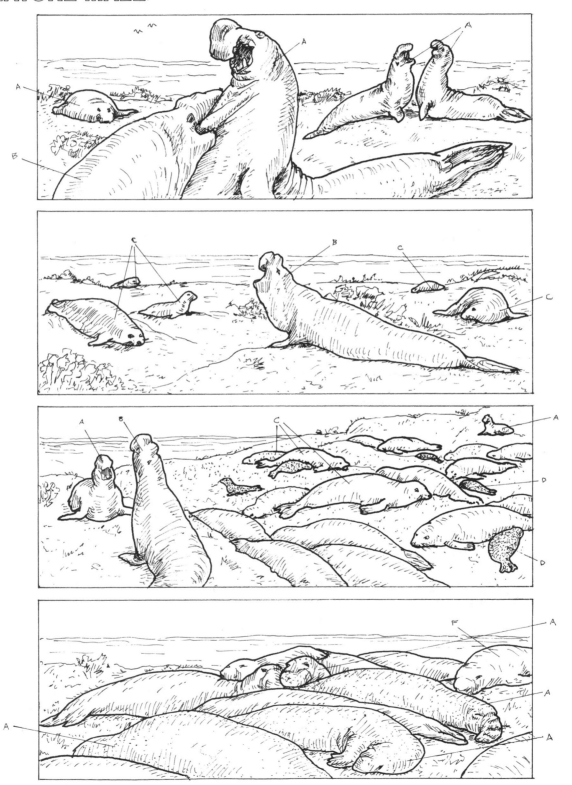

Photosynthetic plants are the basis for the food chains of the sea. The extremely complicated chemical pathways involved in photosynthesis require sunlight for energy, the light-trapping chlorophyll molecule, carbon dioxide (CO_2), water (H_2O), and various inorganic nutrients (chiefly those containing nitrogen and phosphorus). Through these chemical processes, the sun's energy is utilized to convert water and carbon dioxide to oxygen and simple sugars. These products are, of course, used by the plants, but virtually all animals are ultimately dependent on them as well, either directly as herbivores or through the food chain. This plate discusses several marine invertebrates that take shortcuts in the food chain and literally "farm" plants within their own tissues, benefiting directly from the photosynthetic products.

Color each invertebrate farmer as it is mentioned in the text. Begin with the sea slug *Elysia* and the alga, *Codium*, on which it feeds. Note that *Elysia* and *Codium* receive the same color. Next color the single-celled zooxanthellae that are used by the other invertebrates illustrated here. Do not color the green sea anemone located under the rocky ledge.

The bright green sea slug *Elysia viridis* (a gastropod mollusc) is the same color as *Codium*, the green alga on which it feeds. This situation appears at first glance to be a case of substratum mimicry (see Plate 53), but in fact the sea slug's color is acquired from the alga. *Elysia* feeds on *Codium* by slitting open the plant with its radula and sucking in the plant's tissues. Somehow, the green chloroplasts (the photosynthetic organelles of the plant's cells) are transferred intact to the gut lining of the sea slug. These chlorophyll-containing chloroplasts give *Elysia* its green color, but there are additional benefits. The chloroplasts continue to photosynthesize, and the sugars and oxygen they produce are utilized by *Elysia*.

Another mollusc that farms algae is the giant clam (182 kilograms), *Tridacna*, found on shallow coral reefs throughout the eastern tropical Pacific. Like the sea anemones and corals discussed below, *Tridacna* harbors single-celled plants called *zooxanthellae* (modified dinoflagellates, see Plate 12) within special cells in its *mantle*. *Tridacna* is normally found lodged in coral rock, with its hinge directed downward, and its fluted, gaping *shells* facing upward. The thick *mantle*, containing the *zooxanthellae*, projects over the edges of the gaping *shells* in a dazzling display of blue, green, violet, and brown pigments. Besides being beautiful, these pigments serve to shield the *zooxanthellae* from the bright tropical mid-day sun; too much light can inhibit the photosynthetic process. Each night *Tridacna* harvests a portion of its ever-growing plant crop by actually consuming the *zooxanthellae* with special amoebocytes (motile, feeding cells). *Tridacna* still feeds as a normal bivalve filter feeder, but a substantial amount of its nutrition is derived from its crop of *zooxanthellae*.

Zooxanthellae play a significant role in the nutrition of many coelenterates. The giant *green sea anemone* was one of the first coelenterates in which the presence and importance of *zooxanthellae* were discovered. When this *anemone* occupies a habitat in which it is shielded from sunlight, such as in a cave or under a ledge, it is not bright green but is very pale, nearly white. In sunlight, the animal is distinctly green, a color imparted by resident *zooxanthellae*. When a *green anemone* is experimentally transferred to a dark place for several days, the tissues pale dramatically.

The reef-building corals are very much dependent on their partnership with *zooxanthellae*. Not only do these corals receive a substantial portion of their nutrition from the *zooxanthellae*, but they also depend on the microscopic plants to secrete their calcareous skeleton (see Plate 16).

Some corals derive more of their requirements from the *zooxanthellae* than do others. The difference appears to be related to the size of the individual *polyps*. Corals with small *polyps*, such as the *elkhorn coral*, have a high ratio of surface area to volume; this facilitates absorption of sunlight by the *zooxanthellae*. Corals with larger *polyps*, such as the *brain coral*, have a lower surface-to-volume ratio, but the bigger *polyps* are more efficient in trapping zooplankton for food. They still depend upon the *zooxanthellae* in the production of their skeletons.

INVERTEBRATE FARMERS.

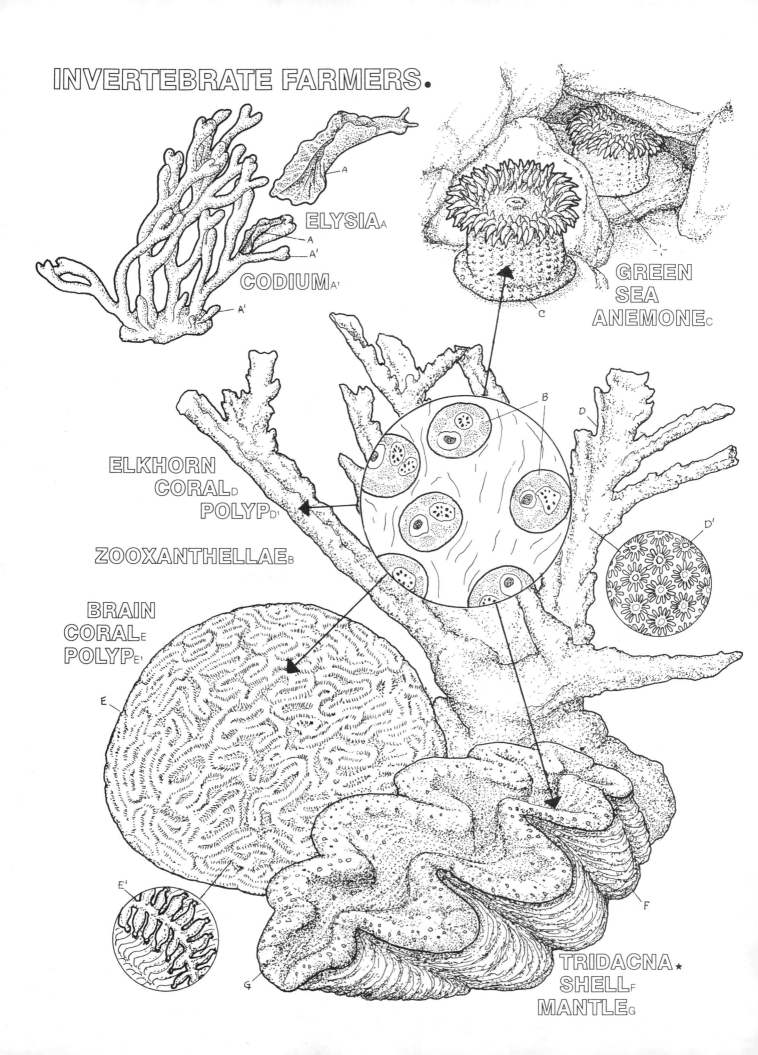

ELYSIA ₐ

CODIUM ₐ₁

GREEN
SEA
ANEMONE ꜀

ELKHORN
CORAL ᴅ
POLYP ᴅ₁

ZOOXANTHELLAE ᴮ

BRAIN
CORAL ᴇ
POLYP ᴇ₁

TRIDACNA ★
SHELL ꜰ
MANTLE ɢ

77

SYMBIOSIS: CLEANER SHRIMP AND FISHES

Many marine fishes are plagued with parasites and diseases, just like other organisms. Small crustaceans (isopods and copepods) and certain worms attach themselves to the outside of the fishes' bodies, onto the gills, or in their mouths. These parasites live at the expense of their hosts and do varying degrees of damage. Bacterial infections also occur on the fins, body surface, and gills of fishes. So common are these parasites and bacteria that some animals specialize in cleaning the affected hosts while feeding on the organisms they pick from the fishes' bodies. Several species of fishes and shrimps engage in cleaning behavior, some examples of which are explained in this plate.

Begin by coloring the illustration of the cleaner shrimp on the anemone at the upper left in the plate; color only the body and antennae. When coloring the enlarged view of the shrimp in the inset, include the claw (D). In the cleaning scene to the right, color only the operculum (F) and mouth (G) of the largest customer fish, leaving the body blank. Color over the entire body of each of the other fish.

Pederson's cleaner shrimp is a common inhabitant of Caribbean coral reefs. This small animal (40 mm) occurs singly or in pairs, and is always associated with a particular species of *sea anemone*. The shrimp either clings to the *anemone* or lives in the same crevice with it; the shrimp derives protection from the *anemone* and is apparently immune to its sting. Many workers believe that each shrimp occupies a *"cleaning station"* in the form of some conspicuous landmark, like a particular coral head. They may station themselves at such points for extended periods of time. The reef fishes apparently recognize and seek out these *cleaning stations* and their tenants when in need of care. When a fish approaches a *cleaning station,* the transparent, violet-spotted shrimp rocks back and forth and whips its long, white *antennae* to signal that it is ready to clean. The fish then swims up and stops within a few inches of the cleaner shrimp, which leaves the protection of its *anemone* and begins to scour the fish's *body* for parasites. The shrimp uses its *claws* to dislodge parasites, and may even make incisions to remove invaders from beneath the fish's skin. When the

shrimp approaches the customer's head, the fish opens each *operculum* (gill cover) and allows the shrimp to enter the gill chamber to clean. The fish then opens its *mouth* for cleaning by the shrimp.

During the actual cleaning process the shrimp is usually safe from predation by its customer fish, even though the fish may be a carnivore that normally feeds on small crustaceans. The reef fishes depend on the cleaners, and, if heavily parasitized or infected, may visit a *cleaning station* several times a day.

Now color the cleaning scene at the bottom of the plate. Color the horizontal bars on both the cleaner wrasse (I) and the imposter blenny (J) black. As in the diagrams above, color only the mouth and operculum of the customer being cleaned, and the entire bodies of the waiting fishes.

Many different types of fishes also exhibit cleaning behavior. In some cases, the young of a species (e.g., butterflyfish and angelfish) are cleaners before they acquire adult feeding habits. These cleaner fishes service a variety of marine animals in addition to bony fishes. Stingrays, mantas, green sea turtles, and even crocodiles have been observed being attended to by cleaner fishes.

One conspicuous full-time cleaner is the small (7–8 cm) *cleaner wrasse* of eastern Pacific coral reefs. This fish is marked with bold, black, horizontal stripes that apparently designate it as a "cleaner" to other fishes. *Cleaner wrasses* often live in pairs near a prominent *cleaning station*. Specific body movements and head-nodding by the *wrasse* serve as an invitation to potential customers. The customer fish then advances and presents itself to the cleaner, often with its *opercula* and *mouth* open. Like the shrimp's clientele, lines of customers may be observed waiting for the services of the *wrasses*.

The relative immunity to predation enjoyed by the *cleaner wrasse* is capitalized on by the *false cleaner blenny*. This small fish has nearly identical markings to the *cleaner wrasse*. However, instead of picking parasites, this fish grabs mouthfuls of healthy tissue from other fishes that have presented themselves to the imposter for cleaning!

CLEANER SHRIMP AND FISHES.

SEA ANEMONE_A
CLEANER SHRIMP★
 BODY_B
 ANTENNA_C
 CLAW_D

CLEANING STATION_E
"CUSTOMERS"★
 OPERCULUM_F
 MOUTH_G
 BODY_H

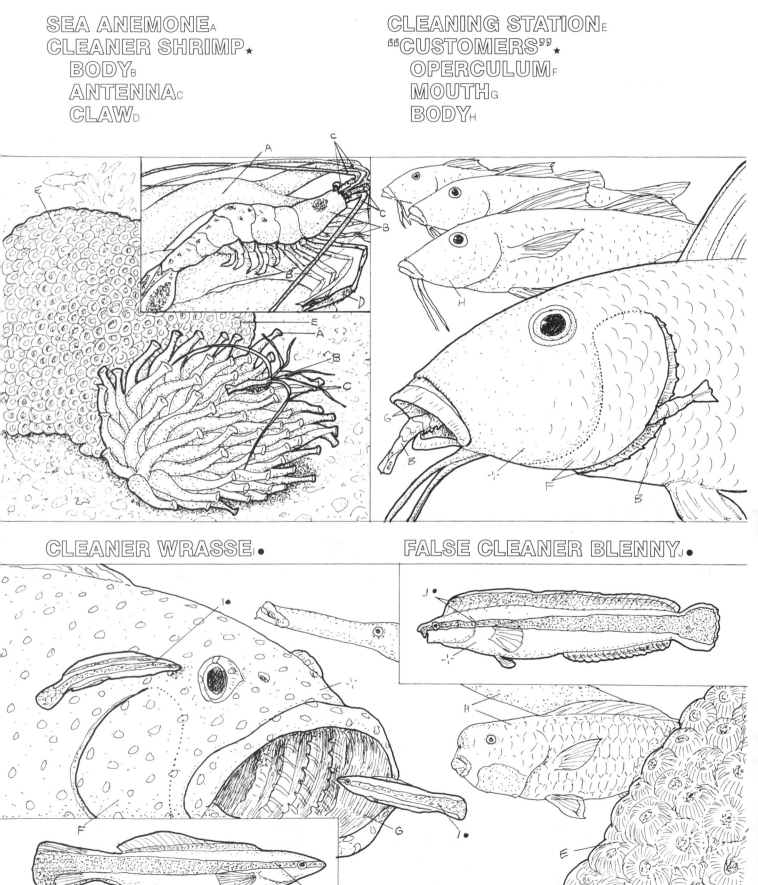

CLEANER WRASSE_I.

FALSE CLEANER BLENNY_J.

78

SYMBIOSIS: ANEMONEFISH AND SEA ANEMONE

The term "symbiosis" refers to the situation in which two dissimilar organisms live together in a mutually beneficial relationship. Such is the arrangement between the *anemonefishes* and sea anemones.

The *clown anemonefishes,* shown here, are small (6–10 cm), orange-colored fish. They are especially abundant on the coral reefs of the western Pacific and Indian oceans where their host anemones are found. The *anemonefish* is dependent upon the protection offered by the sea anemone, and is seldom found on the reef except in the anemone's company.

Color the anemone at the top of the page, and the male and female anemonefishes surrounding it. Next, color the series of four drawings that represent the process of acclimatization of the anemonefish to the anemone. The first two drawings in the series represent the anemonefish's quick withdrawal from the stinging tentacles of the anemone. For a natural effect, leave the white stripes of the anemonefish colorless in these drawings.

Sea anemones are carnivores that subdue their prey by stinging them with the venomous nematocysts on their tentacles. The *anemonefish,* through a process of acclimatization, is able to live among, and be protected by, the anemone without being stung.

Acclimatization may take as little time as a few minutes, or many hours, depending on the species of *anemonefish* and anemone. The fish approaches the anemone and gingerly brushes against its *tentacles* with its tail or ventral surface. It quickly pulls away upon being stung. But the *anemonefish* again returns and gradually brings more of its body into contact with the *tentacles,* until it is able to be engulfed in the *tentacles* with total impunity.

There are several theories regarding the *anemonefish's* acquired immunity to the anemone. One theory suggests that during the process of acclimatization, there is a change in the quality of the mucus coating on the outside of the fish. This change raises the nematocyst's firing threshold, so that when contact is made between fish and anemone, the deadly discharge from the nematocyst is prevented. Another theory suggests that the fish gradually becomes coated by the anemone's mucus, and the anemone does not distinguish the fish from itself, and therefore does not fire its nematocysts.

Now color the family group of anemonefish and their host anemone. Note that the juveniles are of different sizes and slightly different color patterns. Color the butterflyfish which is being challenged by the female anemonefish. Then, color the male anemonefish, the nest, and the eggs in the lower right.

Many species of the *anemonefish* occur in single, tightly structured family groups, living with a single anemone. There is an adult *female* and a *male,* which establish a *nest* site on the substratum adjacent to the *column* of the anemone, within the protective overhang of the anemone's *tentacles.* The two fish clear away all algae and other benthic organisms, and the *female* attaches her *eggs* to the bare substratum. The *male* keeps the *eggs* clean by fanning them with his fins while the *female* continues with the normal feeding behavior of foraging near the anemone for algae, or darting out to grasp zooplankton. The *female* may also exhibit territorial behavior in protecting the host anemone from the anemone-eating *butterflyfish* (as shown in the illustration).

The *eggs* of the *anemonefish* hatch in 6 to 8 days, and the larvae live several weeks in the plankton. Once settled, the small fish remain among the *tentacles* of the host anemone, feeding directly on the *tentacles* or on material egested by the anemone. When a small fish ventures away from the anemone, it is noticed and chased by the resident adults and other *juveniles.* Many *juveniles* are chased away in this manner, and among those that remain in the original group, a dominance hierarchy is established. *Juveniles* that leave the original group must find a different anemone with which to live, and go through the acclimatization process with this new host.

The *anemonefish* constantly re-establishes contact with the anemone. During feeding forays or nest-tending, the fish will frequently return to brush against the anemone's *tentacles.*

SYMBIOSIS: ANEMONEFISH AND SEA ANEMONE.

ACCLIMATIZATION ★
  SEA ANEMONE ★
    TENTACLE A
    COLUMN B
MALE ANEMONEFISH C

FEMALE ANEMONEFISH D   JUVENILE G
NEST E   BUTTERFLYFISH H
  EGG F

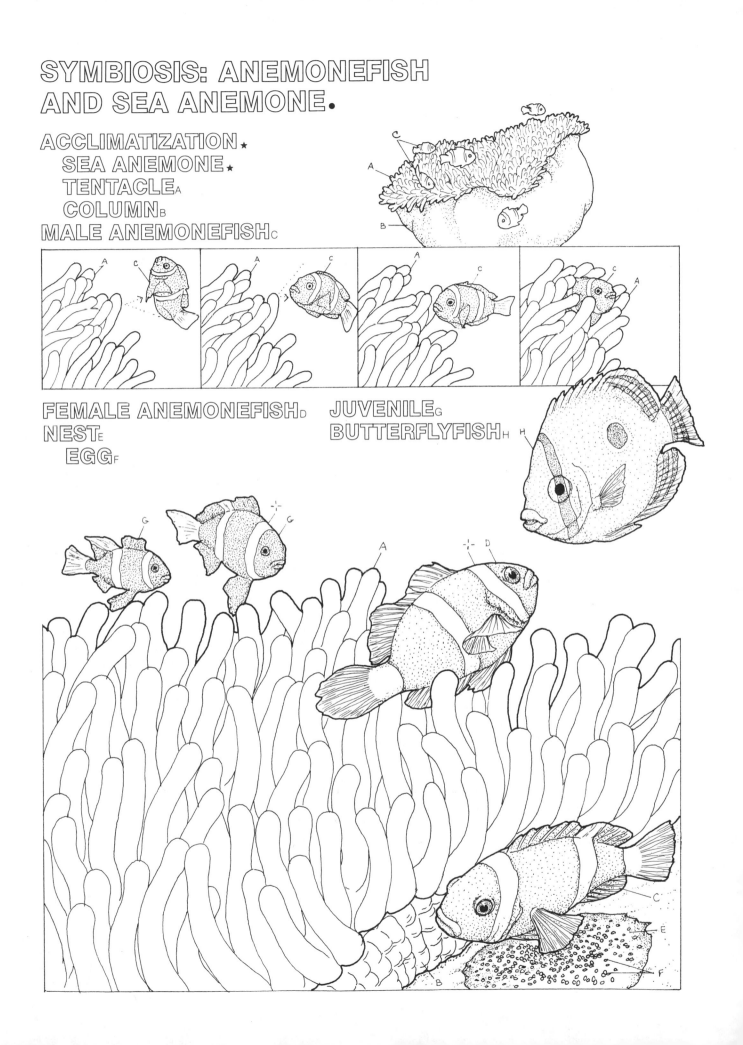

79

INTRASPECIFIC AGGRESSION IN SEA ANEMONES: CLONE WARS

Color the uppermost sea anemone and its tentacles, oral disc, body, and pedal disc. The anemone is undergoing asexual reproduction. Next, color the two sea anemone clones (C¹ and G) in the rectangle, noting that clone C¹ is derived from the continued asexual reproduction of the anemone at the top of the page, and receives the same body color. The anemone-free area between clones should be colored gray.

The small aggregating sea anemone, commonly found in the middle rocky intertidal zone of the Pacific coast, reproduces asexually by binary fission. The *pedal disc* moves in two directions at once, stretching the sea anemone and ultimately pulling it in half. Each half becomes a new anemone, which can grow and split again. If the sea anemone is in a favorable spot with abundant food and room to spread, it will continue to reproduce asexually, and soon a group of identical sea anemones is established. Since all the anemones originated from the same individual, each anemone possesses exactly the same genetic material; the group as a whole is known as a genetic *clone*.

Different *clones* in the same locale can be recognized by distinct color patterns on the tentacles and oral discs, which are specific to each *clone*. When two different *clones* are found in a given area, the interface is often marked by a patch of bare rock, called an *anemone-free area*. Sometimes barnacles or other animals inhabit these spaces, but both anemone *clones* avoid it. Laboratory experiments conducted by Lisbeth Francis at the Univeristy of California at Santa Barbara provide our best understanding of the *anemone-free area* and the *clone* wars.

Color the four stages of aggressive encounters. The two anemones are members of different clones. Note that the acrorhagi (E) and the ectodermal tips on clone C¹ are each colored a different color than the body. Note that in the illustration of withdrawal the ectodermal tips end up on the body of the victim clone.

When the tentacles of two members of the same *clone* come in contact, both anemones initially contract. Then they gradually extend their tentacles so that they interlace, and the two anemones have no further interaction.

If two anemones from different *clones* have *tentacle* contact, they also initially retract. This expansion and retraction of the *tentacles* may occur several times. Eventually, one of the anemones begins to inflate special structures called *acrorhagi,* which are located just beneath the oral disc. Water is forced into the hollow *acrorhagi,* and they become distended and cone shaped. The anemone may also rise up by contracting the circular muscles in its column or body, and become inflated. The towering aggressor then sweeps down with extended *acrorhagi* and attacks the alien anemone. The *acrorhagi* have large, penetrating nematocysts in their *ectodermal* tips that discharge into the victim.

This attack may be repeated by the aggressor anemone, or the victim may retaliate with similar behavior. More often, however, the victim pulls its body away from the aggressor in an attempt to avoid further contact. If there is no room to move back, the victim may release its hold on the substratum and be washed away by the water. If there is absolutely no avenue of escape, and the victim is left in contact with the non-clone-mates, it may be killed in a matter of days.

This interaction between *clones* explains the existence of *anemone-free areas* between adjacent *clones* in the natural habitat. Those anemones on the leading edges of interacting *clones* pull back to avoid the mutually aggressive encounters, and an open area remains.

CLONE WARS.

SEA ANEMONE★
TENTACLES$_A$
ORAL DISC$_B$
BODY$_C$
PEDAL DISC$_D$

CLONE 1$_{C^1}$
ACRORHAGI$_E$
ECTODERM$_F$
CLONE 2$_G$

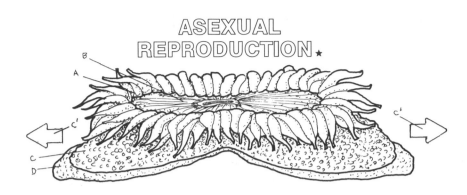

ASEXUAL
REPRODUCTION★

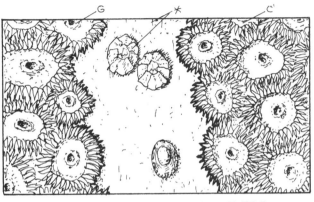

ANEMONE-FREE AREA★

A. CONTACT★

B. INFLATION★

C. ATTACK★

D. WITHDRAWAL★

The acorn barnacle is a familiar marine organism. Although sessile as an adult, it is first a swimmer and then a crawler in the larval stage. Thus, the choice of attachment site by a settling *cyprid larva* is literally a life or death decision for the individual barnacle—it could settle in an area where no food is available; and for the species—reproduction requires close proximity of members of the same species (see Plate 66). Barnacle species, therefore, have evolved complex mechanisms to insure successful and neighborly settlement. In this plate, how such mechanisms operate in a single species and how settlement affects interactions between barnacles competing for space will be explored.

Begin by coloring the illustration of settling behavior at the top of the page. Note that the path of movement of the crawling cyprid larva receives the larval color (A) as does the attachment site. The adult and the basal plates previously occupied by adults also receive the same color (B).

The settling mechanism of the common Atlantic intertidal barnacle *Balanus balanoides* was studied by Dennis Crisp and others. They have demonstrated that the *cyprid larva* of this barnacle can recognize a proteinaceous substance that is found on the *adult* barnacle's body as well as the *basal plate,* which is left behind if the *adult* barnacle is removed. When the crawling *cyprid larva* contacts this protein substance with its antennae, it begins crawling in a circular pattern. As long as it intercepts *adult* barnacles of its own species it continues to crawl. Once it encounters a clear area, it tightens up its circular pattern to make sure the area is large enough for adequate growth. If the site is acceptable, the *cyprid* attaches (see Plate 66). This settling behavior insures that the selected area is suitable for feeding and reproduction (indicated by the presence of other adults).

Color the two rectangular drawings; note the size of the young barnacles relative to the thumb in the drawing. Note that the pencil-form adults

and previously occupied basal plates receive the same color (D).

In *Balanus glandula,* a U.S. Pacific coast intertidal species, this gregarious settling behavior may result in overcrowding. A collection of young barnacles *(spat)* may settle out very close to one another, suggesting a capacity for recognition quite similar to *Balanus balanoides.* However, the spacing between individuals in areas of particularly dense settlement is insufficient for lateral growth. The barnacles have to grow upward to reach full size. This results in a *pencil-form adult* instead of the typical volcano-shaped barnacle. This *pencil form* is more easily dislodged by predators or floating debris, leaving vacant its *basal plate.*

Color the two examples of interspecific settlement. Note that the arrows receive the Balanus balanoides' adult color in the bottom example.

Settling behavior by *Balanus* barnacles does not include recognition of species other than its own. If there are no adults of its own species present, the cyprid selects a coarse or rough-textured substratum for settlement. In the example shown here, *spat* of *Balanus glandula* have chosen the rough shell of the much larger thatched barnacle, *Tetraclita squamosa.*

Another, perhaps more dramatic, example of interspecific indifference was discovered by Joseph Connell. The Atlantic species *Balanus balanoides* and *Chthamalus stellatus* both grow successfully in the high rocky intertidal zone in Scotland. Although they both settle in groups of mixed species, the more rapidly growing and larger *Balanus* spaces itself as described above but does not recognize the presence of the smaller *Chthamalus.* This results in *Chthamalus* being: 1) overgrown and smothered; 2) uprooted by the growing edge of a *Balanus;* or 3) literally squeezed to death by the rapid growth of several adjacent *Balanus.* In this manner, *Balanus* effectively eliminates *Chthamalus* from the substratum on which they both reside. *Chthamalus,* as a species, is able to persevere in the face of this competition because it has a greater tolerance to exposure and can live higher in the intertidal zone than *Balanus.*

BARNACLE INTERACTION.

**BALANUS BALANOIDES
SETTLING BEHAVIOR ★**
CYPRID LARVA_A
   MOVEMENT_{A1}
    ATTACHMENT SITE_{A2}
ADULT_B
   BASAL PLATE_{B1}

**OVERCROWDING IN
BALANUS GLANDULA ★**
SPAT_C
PENCIL-FORM ADULTS_D
   BASAL PLATES_{D1}

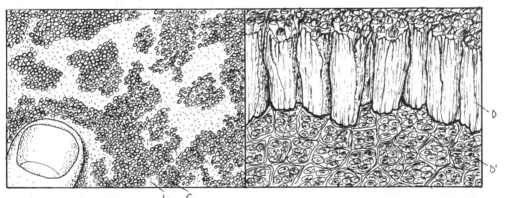

INTERSPECIFIC SETTLEMENT ★
TETRACLITA_E
BALANUS GLANDULA_{C1}

BALANUS BALANOIDES ★
CHTHAMALUS_F

SMOTHERING ★   UPROOTING ★   SQUEEZING ★

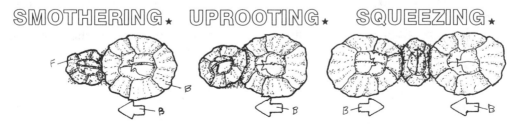

In marine environments where a single resource is highly contested, complex relationships will often develop among competing organisms. One of the most fascinating examples of such a relationship involves the interaction of the large Pacific coast *owl limpet*, with its intertidal competitors for a primary resource: space. This *limpet* was studied by John Stimpson at Santa Barbara, California. It can grow up to 8 centimeters long and is most commonly found in the mid-littoral zone of the rocky intertidal area, where strong wave action occurs. This subhabitat is occupied by *mussels* and *stalked barnacles* which usually cover all open space and crowd other organisms. In such a situation, it would at first seem surprising to find bare patches in the middle of a dense *mussel* and *barnacle* clump (see Plate 95). Closer inspection usually reveals the presence of a single, large *limpet* within each clean patch. Stimpson determined that these resident *limpets* literally "farm" these patches and defend them from encroaching competitors.

Begin by coloring the limpet in the upper drawing. This is its actual size. Its shell would be a dark gray or light brown. Now color the limpet on its algal turf (M), which should be colored a light green or yellow-green. In the lower right-hand magnified view, color the whole circle with this light color, then use a darker color to fill in the radular scrapings (E¹).

The *limpet* patches average about 900 square centimeters (one square foot) in area and their yellow-green color sharply contrasts with the bare rock areas adjacent to the patch. The yellow-green color is due to a low-growing *algal turf* (about 1 mm in height) composed mainly of filamentous blue-green algae on which the *limpet's* distinctive *radular scrapings* can be seen (see Plate 89). These *radular scrapings* are rectangular and are about 1 by 3 millimeters in size. As these large *limpets* graze on their patches, they leave behind a coarsely cropped *algal turf*. The radulas of smaller limpets (less than 20 mm) rasp off the *algal turf* much closer to the substratum and leave behind what appears to be bare rock. With each high tide the grazing *limpet* leaves its roosting place, called a *home*

scar, and moves in rotation to a different portion of its patch. By this time the algae have sufficiently regrown in the earlier grazed portion, so that the *limpet* can begin the grazing cycle all over again in much the same way that a farmer rotates his cattle through a series of separate pastures. It was found during four years of observation that many *limpets* stayed on the same patch throughout that period, which is a relatively long time in such a wave-tossed habitat.

Now color the mussels, stalked barnacles and the predatory snail both on the algal turf and in the peripheral drawings. If you wish to use naturalistic colors, the mussels would be blue, the barnacles light gray or tan, and the predatory snail brown or orange.

If the patch-dwelling *limpet* is removed, and smaller limpets allowed to graze within the patch, they will graze the *algal turf* down to bare rock in two weeks, leaving nothing for the large *limpet* to farm. Also the *mussels* and *barnacles* that often form the borders of *limpet* patches are constantly encroaching on the patch. In response to these competitors, the *limpet* shows a dazzling repertoire of behaviors. When confronted with another limpet (not shown) on its patch, the resident *owl limpet* backs up and rams its *shell* into the intruder, repeatedly, until it either loses its grip on the substratum and falls off, or retreats from the patch. To discourage encroaching *mussels* and *barnacles*, the *limpet* uses its large *shell* in bulldozer fashion to push the invaders back, probably dislodging the mussel's *byssal threads* and undercutting the barnacle's *basal attachment*. The farmer *limpet* saves its most spectacular behavior for predatory gastropods that may venture onto its patch. When a *snail*, which is an occasional predator of small limpets, is encountered, the *limpet* (putting aside its battering method) raises its *shell* high off the substratum, in a sort of wind-up, and brings the leading edge quickly down on the front of the predator's foot. The predator responds by releasing its grip on the substratum in an effort to retreat into its shell, and the *limpet* then eases up, allowing the predator to be washed away by the surf.

THE FARMER LIMPET.

OWL LIMPET_A
SHELL_{A'}
FOOT_B
HEAD_C
   MOUTH_D
   RADULA_E
MANTLE_F

MUSSEL_G
   BYSSAL THREAD_H
STALKED BARNACLE_I
   BASAL ATTACHMENT_J
PREDATORY SNAIL_K
   BODY_L
ALGAL TURF_M
   RADULAR SCRAPING_{E'}
   HOME SCAR_N

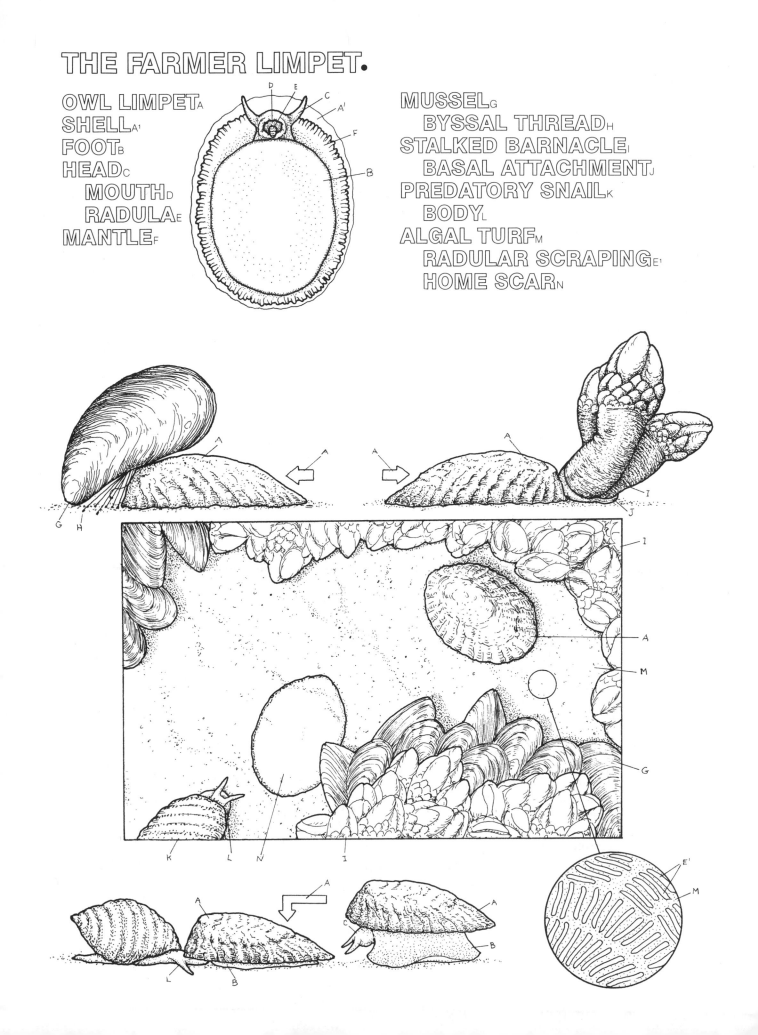

82

SEA PALM STRATEGY

In the rocky intertidal there is constant competition for space, a competition in which plants are set against sessile animals. Both the California sea *mussel* and the *sea palm,* a brown alga, occupy mid-intertidal areas of heavy surf in the Pacific Northwest. The *mussel* is a conspicuous and dominant organism that forms extensive beds. *Mussels* are both secured to the rocky substratum and linked to each other by their tough byssal threads (see Plates 23 and 95). How an annual plant (living only one year) like the *sea palm* is able to persist in this environment against such competition involves an interesting interplay of biological and physical factors.

Color the sea palm at the upper left of the plate. Next color the scene at the bottom of the page, noting that the holdfast (C) of the beached plant contains the remains of barnacles (F) and a piece of an algal colonizer (G). Finally, color the rectangular diagrams; note that only one color is used for the entire sea palm.

Sea palms appear in February or March of the year and grow rapidly from April to June. They persist through the summer but die back in autumn; they are gone by the end of November. The individual *sea palm* has a large *holdfast* that anchors it in the wave-swept habitat. The *stipe* tapers upward from the *holdfast* and is crowned with many *blades* that give the alga its palmlike appearance. And like the true palm, whose flexible trunk sways in the tropical winds, the *sea palm's stipe* bends and returns after each crashing of a wave.

The *sea palm* rarely occurs singly; it is usually present as a tight clump of various-sized individuals. This clumping occurs partly because the plant releases its *spores* during low tides in the spring and summer, causing many *spores* to land nearby and grow up on the *holdfast* of the adult plant. This spore-releasing has important consequences.

The expansive *mussel* beds are vulnerable to predation by the sea star (see Plate 95) and to the impact of large, wave-borne objects like logs. Logs bash into the *mussels,* crushing and dislodging individuals and exposing the bed to further erosion by wave action. Large areas of substratum are cleared in this way, and many species of *algae* and barnacles quickly *colonize* the newly opened space. If there is a *sea palm* clump nearby to serve as a source for *spores* or, if a spore-bearing portion of the *sea palm* drifts on to this space, young *sea palms* will grow.

Sea palms may themselves help to bring about clearing of the substratum. They are able to establish themselves and grow on top of some other organisms, including algae and *barnacles*. The *sea palm's holdfast* provides the usual secure anchorage on the underlying organism, and, as the *holdfast* grows, smothers it. While this is occurring, the *sea palm's stipe* is likewise growing and absorbing more and more impact of the breaking waves. The plant is well able to bear this stress, but the organism to which its *holdfast* is fixed is now dead, decaying, or increasingly unstable; eventually the whole mass is torn off, leaving the bare rock.

It commonly happens that the *sea palm*—the conspicuous sporophyte plant—is torn away late in the season by early fall storms. The rocky substratum may appear to remain bare through the winter, but it is not unlikely that the tiny *sea palm* gametophytes have colonized the area and will provide for the large sporophyte plant to establish itself there the following spring (see Plate 58).

In this manner, the clump of *sea palm* can replace itself annually and persist in an area for several years until the slow, steady growth of the *mussel* bed overtakes it. The presence of the *sea palm* in an area is dependent on a nearby source of *spores* and on fairly frequent interruptions of the *mussels'* monopoly on space, either by physical or biological agents (wave-borne debris or sea star predation).

SEA PALM STRATEGY.

SEA PALM A
BLADE A1
STIPE B
HOLDFAST C
SPORE D
MUSSEL E
BARNACLE F
ALGAL COLONIZER G
SEA STAR H

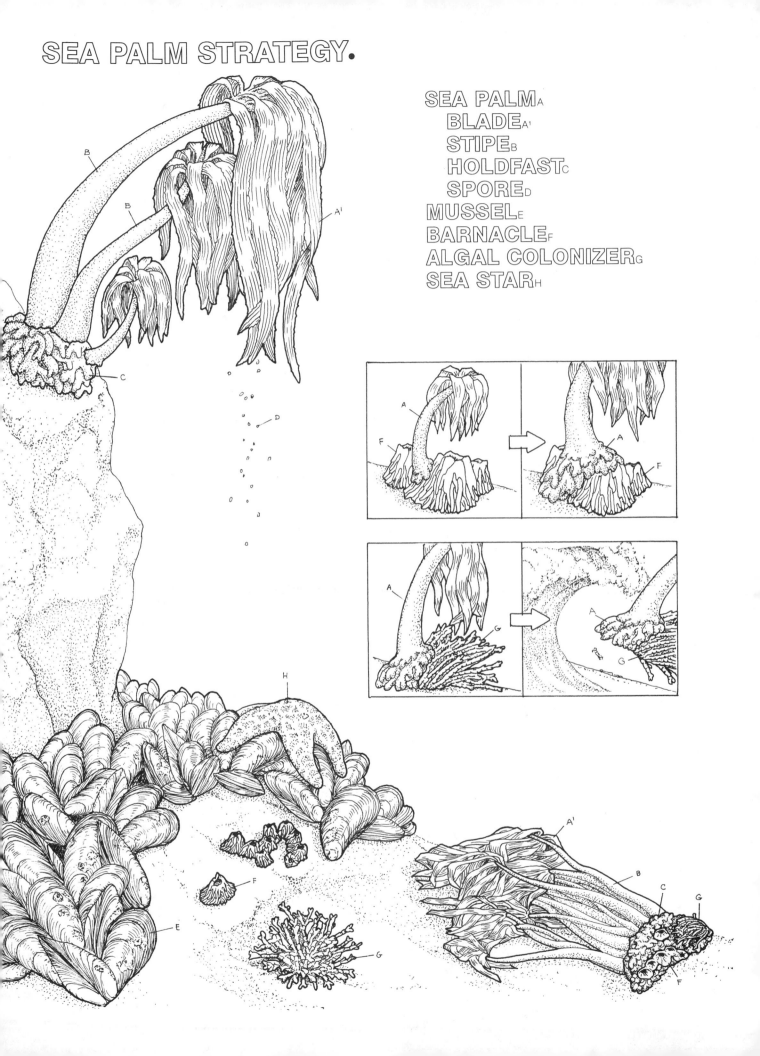

83
DEFENSE MECHANISMS: THE NEMATOCYST POACHER

Aeolid nudibranchs feed primarily on coelenterates, especially hydroids and sea anemones. Their ability to eat these organisms without themselves succumbing to the coelenterate's *nematocysts* is quite remarkable. In addition, they not only eat coelenterates and their *nematocyst*-studded *tentacles,* but these nematocysts are utilized by the nudibranchs for their own protection. This plate illustrates how the nudibranch accomplishes this task.

Color only the tentacles of the sea anemone at the top left. Color both parts of the arrow (B) suggesting the swallowing of nematocysts from the tentacles to the gut of the nudibranch. Color the cerata of the nudibranch (in life they are often a pale pink). Then go to the diagram of the nudibranch gut on the left. After completing the gut, color the enlargement of the cnidosac (F), in the lower left corner, and the coiled and discharged nematocysts to its right. You may wish to use gray for the fish predator, shown being warded off by the discharge of nematocysts from the nudibranch's cerata.

Aeolid nudibranchs are relatively common in the lower intertidal and subtidal zones in cold temperate waters along both the Atlantic and Pacific coasts. On the Pacific coast, some species feed on sea anemones, including the aggregating anemone. The nudibranch approaches the anemone and, after initial contact, pulls back, erecting the *cerata* forward over the head. It then moves in to feed. By pulling its body in under the *cerata,* it resembles another anemone. The nudibranch prefers to feed on the anemone's *tentacles,* and bites off pieces with stout bladelike jaws (not shown). When these *nematocyst*-bearing portions are ingested, the *nematocysts* are swallowed intact and somehow remain undischarged. How the nudibranch prevents *nematocyst* discharge is a mystery; it may possibly secrete a special mucous substance that immobilizes the discharge mechanism. Once swallowed, the *nematocysts* are moved along special ciliary tracts from the *gut* (stomach) into *gut diverticula* (sacs), located in each of the nudibranch's *cerata*. Aeolid nudibranchs use these *gut diverticula* to digest their food. However, the *nematocysts* are not digested, instead they are moved to the tips of the *cerata*. Here the undischarged *nematocysts* are maintained in special *cnidosac cells,* located within larger pouches called *cnidosacs* that open to the outside.

When the *cerata* are roughly touched or pulled off the nudibranch, as would occur when a predatory fish moved in for a meal, special circular muscles around the *cnidosac* contract and expel the *nematocysts*. These promptly *discharge,* often into the mouth of the predator, embedding themselves in its tongue and nearby soft tissues. One or two of these encounters will cause a fish to delete these nudibranchs from its diet!

The process involved in this defense mechanism is not fully understood. When the *cnidosacs* empty their contents, new *nematocysts* are replaced in three to twelve days. Most aeolid nudibranchs employ only certain types of *nematocysts* from among the several kinds present in their coelenterate prey.

THE NEMATOCYST POACHER.

SEA ANEMONE ★
TENTACLE A
    NEMATOCYST B

NUDIBRANCH ★
GUT C
CERATA D
    GUT DIVERTICULA E
    CNIDOSAC F
        CNIDOSAC CELL G
        NEMATOCYST
        (DISCHARGED) B¹

PREDATOR
FISH ★

MARINE INVERTEBRATE DEFENSE RESPONSES

The study of marine invertebrate defensive responses has revealed some interesting patterns. Some species will react to local predators but not to previously unencountered ones, though they may be closely related. Within a single group, some invertebrate species will show very similar defensive behaviors, while others lack them entirely. Still more intriguing are the parallel behaviors that have evolved in some species belonging to widely separated groups. In this and the next three plates we will investigate invertebrate defense behavior.

Color the keyhole limpet and its predator, the ochre sea star. The drawing on the far upper left depicts the normal appearance of the limpet, while the upper middle drawing shows the full response to the sea star.

Sea stars occur on most ocean bottoms and feed on a wide variety of invertebrate prey. The keyhole limpet, when touched by the *tube feet* of the common, intertidal *ochre star*, shows a quick and effective response. The limpet pushes its *shell* upward and throws one fold of its *mantle* up around its *shell*, and another fold down around its *foot*. It also erects its *siphon*, which overlaps the top of the *shell*. This behavior effectively covers the *shell* with soft tissue that is difficult for the *ochre star* to grip. It appears that the *ochre star* may seek to avoid the limpet's *mantle*, as contact between the *mantle* and the sea star's *tube feet*, when they are on the limpet's *shell*, causes the *star* to release its suction hold. The keyhole limpet's *mantle* may secrete some substance repulsive to the *ochre star*.

Color the sea urchin–leather star interaction in the middle of the plate. Note that the star has been repulsed by the urchin and still has pedicellariae (J) clinging to its tube feet. The drawing on the far left is a close-up of the urchin's surface.

Looking at the pin-cushion appearance a *sea urchin* puts forward, one wonders how anything could eat it. However, a number of sea stars will readily devour *sea urchins* if given the chance. One such sea star is the *leather star* (of the west coast of the United States) which feeds on the purple *sea urchin,* especially in shallow water off southern California. If the *leather star* approaches the *sea urchin* on a level surface, two responses can occur. First, the *urchin* may simply move away quite rapidly; urchins have been observed moving three times their normal speed after encountering the *tube feet* of predatory sea stars. The second response involves the special stalked appendages known as *pedicellariae,* of which there may be two or more types. *Pedicellariae* are found all over the *urchin's* body and possess three stout, articulating jaws (not shown). When approached by a predatory *leather star,* the *urchin* pulls in its *tube feet,* lowers its *spines,* and erects batteries of *pedicellariae* that grab onto and pinch the skin of the *star.* In some cases, poison glands may secrete their fluid into the wound made by the jaws of the *pedicellariae.* This counterattack is often sufficient to discourage the *leather star,* which retreats with its *tube feet* festooned with the *urchin's* uprooted, but tenacious, *pedicellariae.*

Color the bottom illustrations, which demonstrate the defense responses of the sea anemone and feather-duster worm. The drawing at left shows their undisturbed appearance; their defensive postures are shown at right.

A more direct, but less colorful mechanism is seen in sea anemones and feather-duster worms. Both of these animals simply hide when danger comes near. A fully erect sea anemone, when disturbed, will retract its *tentacles* into its central cavity; if provoked further, it will expel all the sea water inside and pull down against the substratum (not shown). The feather-duster worms quickly retract their *tentacles* into their calcium carbonate *tubes* whenever touched. Light-sensitive receptors on their *tentacles* cause these worms to retract when a shadow passes over or when light is shined directly at them. Feather-duster worms seldom hide inside their *tubes* for very long, and soon their crown of filter-feeding *tentacles* is spread open again.

MARINE INVERTEBRATE DEFENSE RESPONSES.

KEYHOLE LIMPET★
SHELLA
HEADB
MANTLEC
SIPHOND
FOOTE

OCHRE STARF
TUBE FOOTG

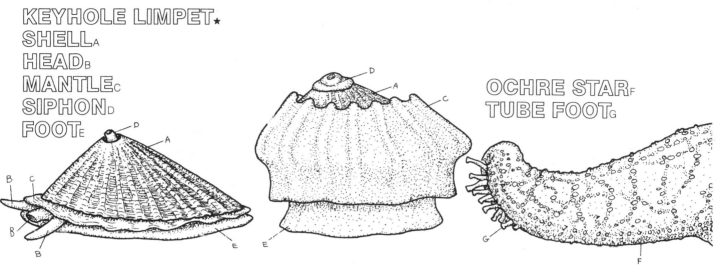

SEA URCHINH
SPINESH'
TUBE FOOTI
PEDICELLARIAJ
POISONOUS
   PEDICELLARIAJ1

LEATHER
STARK

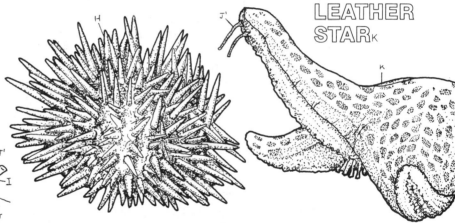

FEATHER-DUSTER WORM★
TUBEL
TENTACLEM

SEA ANEMONE★
BODYN
TENTACLEO

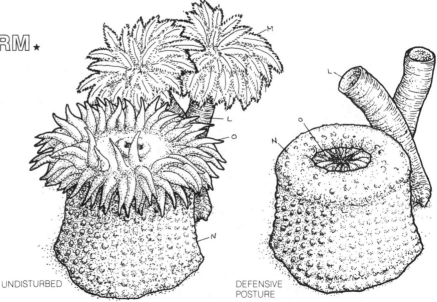

UNDISTURBED

DEFENSIVE
POSTURE

The defense mechanisms of fish often involve the use of spines. We have already seen the poisonous spines of the lionfish (Plate 49), stonefish (Plate 50), and stargazer (Plate 37), which provide a deadly and effective deterrent. Spines are utilized in a variety of ingenious mechanisms, as this plate illustrates.

Begin with the normal (lower view) and inflated (upper view) porcupine fish. In the inflated porcupine fish (with erect spines), note that the spines and body both receive the color used for the spines; a light color is recommended. In the lower illustration of the porcupine fish, the spines lie flat; color the body and spines the color chosen for the body (A); this should also be a light color.

The first, and by far the most spiny fish to be considered is the porcupine fish. This fish is cosmopolitan in warm waters and can grow quite large (about 1 meter). Under normal circumstances the porcupine fish goes about its business of hunting among the crevices of coral reefs for molluscs, crustaceans, and echinoderms that it crushes with its stout, beaklike *jaws.* Its protective *spines* are folded nearly flat with their tips facing toward the back. When molested or threatened, the porcupine fish rapidly inflates its *body* by swallowing water and becomes nearly spherical in shape. The once-flattened *spines* now become fully erect and stick out in a menacing, pin-cushion fashion. Once the danger has passed, the porcupine fish quickly expels the water and resumes its normal activities. The erect *spines* can cause serious puncture wounds. Among South Pacific Islanders, the spiny skins of porcupine fish were once used for helmets.

Color the surgeonfish and the inset magnified view of the hinged spine protruding from its body wall.

The surgeonfish is another familiar resident of coral reefs. This medium-sized (15–60 cm) fish nibbles on filamentous algae with its small terminal *mouth.* When bothered by an intruder, this peaceful herbivore erects a pair of lancelike *spines* located on either side of the base of the tail. The *spines* are actually pairs of modified scales hinged on the posterior end, so that their sharp inner edge faces forward when erect. When not in use the *spines* retract into horizontal grooves. The *spines* are razor sharp and have inflicted many a nasty wound on unsuspecting fishermen. They may possibly be used to slash other fish, or

perhaps only as a threat display. Many surgeonfish species have the *spines* boldly outlined in a color that contrasts with the surrounding body color. A warning sweep of the tail, flashing these colors and the erect *spines* often deters a would-be transgressor.

Color the triggerfish and note that the spines under discussion are actually part of the dorsal fin. Color these spines the color you chose for the spine (I), and color the rest of the fin the color of the dorsal fin (B). The inset above the fish shows the spines folded. Be careful to leave the outlined spots of the triggerfish white to maintain the high contrast coloring of this fish.

When trouble courts the clown triggerfish (25–50 cm) of the tropical Pacific it quickly finds shelter in a small cave or crevice in the coral reef. Once there, the triggerfish erects the long first *spine* of its *dorsal fin* and locks it into place from behind with the smaller second dorsal *spine,* thus wedging itself tightly in place. The triggerfish cannot be removed from its cave without breaking the stout, dorsal *spine.* To fold the first *spine* backward, the smaller second *spine* (the "trigger") must be depressed. The triggerfish will also raise its *dorsal fin* in any situation of attack or defense. This presents a more formidable posture and makes the fish look bigger.

Now color the shrimpfish and the spines of the sea urchin.

The last spiny trick considered here is most unusual. The shrimpfish is a small (15 cm) inhabitant of coral reefs and sea grass beds of the tropical Pacific. Shrimpfish have a long snout and are very highly compressed laterally. They are encased in a transparent armor consisting of separate plates that taper rearward into a long point which is the first spine of the *dorsal fin.* Shrimpfish swim by undulations of their fins and always in a vertical position, head down. When feeding on small invertebrates—and especially when it is not moving—this long, thin shape is hard to distinguish from the blades of sea grass. On the coral reef the shrimpfish borrows the spiny protection of long-spined sea urchins. The small size and elongated form of the shrimpfish allows it to fit comfortably among the *spines*—a relationship that deters most predators. This spiny trick is not unique to the shrimpfish, but its shape and the long, black stripe along its body make urchin *spines* an especially effective hiding place for this species.

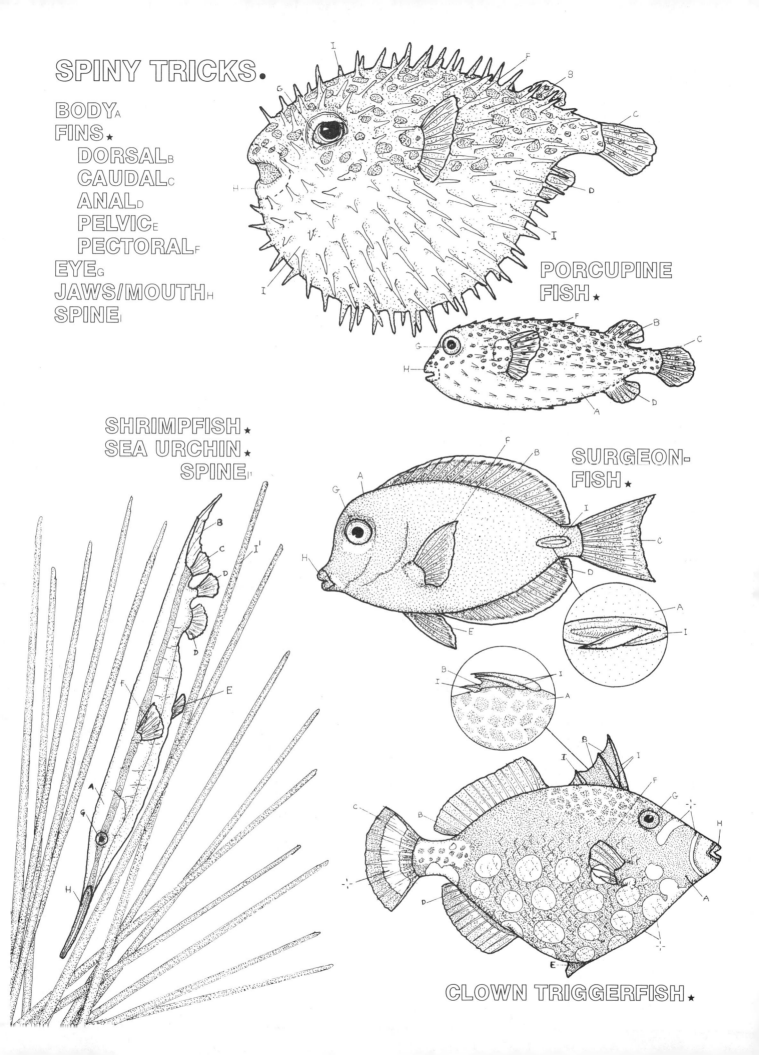

SPINY TRICKS.

BODY A
FINS ★
   DORSAL B
   CAUDAL C
   ANAL D
   PELVIC E
   PECTORAL F
EYE G
JAWS/MOUTH H
SPINE I

PORCUPINE FISH ★

SHRIMPFISH ★
SEA URCHIN ★
   SPINE I¹

SURGEON-FISH ★

CLOWN TRIGGERFISH ★

86
DEFENSE MECHANISMS: SAYING NO TO A SEA STAR

Sea stars are unlikely-looking predators. Seen at the beach when the tide is out, they often seem not to move at all. Even when watched underwater, their movements seem slow and without menace. However, when these movements are speeded up by time-lapse photography, the predatory skills of the sea star become quite apparent. The foraging sea star is a relentless, multi-armed predator that slowly but surely overpowers its prey and immobilizes them with its many tube feet. A number of the more mobile marine invertebrates have developed modes of behavior that often allow them to escape sea star predation. In this plate we will see three ways invertebrates say "no" to a sea star.

Color each of the three escaping invertebrates and the related sea star predator, commencing with the scallop and ending with the anemone. Note that the short-spined star (A) is a predator of both the scallop and the cockle, and will trigger an escape response in each animal.

The bivalves known as scallops are not, as one might suspect, sedentary invertebrates. The posterior adductor muscle (the edible part of scallops), which is used in other bivalves to close the shell and keep it closed (see Plate 22), is centrally located in the scallop *shell.* It is able to rapidly contract, expelling water out of the scallop. The *mantle* is muscular along the edge and forms openings through which the expelled water is directed out in propulsive *jets,* sending the scallop through the water in the opposite direction. When swimming normally the scallop moves with the hinge (straight edge of shell) facing backward and appears to be "chewing" its way through the water with rapid clacking of its *shells.* When a predatory sea star, like the *short-spined star,* touches a resting scallop, the startled mollusc leaps from the substratum propelled by *jets of water* forced out through the front of its shell, as illustrated. The scallop may leap a meter or more, putting it well out of the sea star's grasp.

This escape response isn't foolproof; the scallop may jump straight up and come down right back on top of the sea star.

Another molluscan high-jumper is the Pacific coast cockle. The cockle burrows just below the substratum surface and has a large digging *foot* to reburrow itself if dislodged. If touched on the *mantle, foot,* or siphons (not shown) by a predatory sea star, the cockle responds swiftly. The *foot* extends out from the *shell* and curls back underneath it; then with a powerful thrust, the muscular *foot* extends its full length away from the *shell,* giving it a tremendous push against the substratum. In this way, the cockle is vaulted several centimeters into the water and may continue for several leaps until it is well out of harm's way. If the cockle is touched by one's finger or other non-sea star material, it will only clamp its valves shut and remain in place. So far as is known, the escape response just described is triggered only by the cockle's sea star predators.

Although sea anemones are thought of as fairly sedentary types, almost all can move about to some extent by a slow creeping movement of the *pedal disc.* Such movement is measured in centimeters per hour. However, one sea anemone of the Pacific Northwest shows a very rapid escape response when confronted by the *leather star,* a known predator of the sea anemone. When touched by the *leather star,* the anemone withdraws its *tentacles* while at the same time detaching its *pedal disc* from the rock substratum. In a few seconds, the anemone is completely detached and a cone-shaped projection is visible in the center of the *pedal disc.* The anemone then begins an awkward but effective swimming movement away from the *leather star* by the rapid back-and-forth bending of its column-shaped *body.* The whole escape takes little more than ten seconds! After moving some distance, the anemone will sit quietly for awhile and, if unmolested, will reattach itself in about 15 minutes.

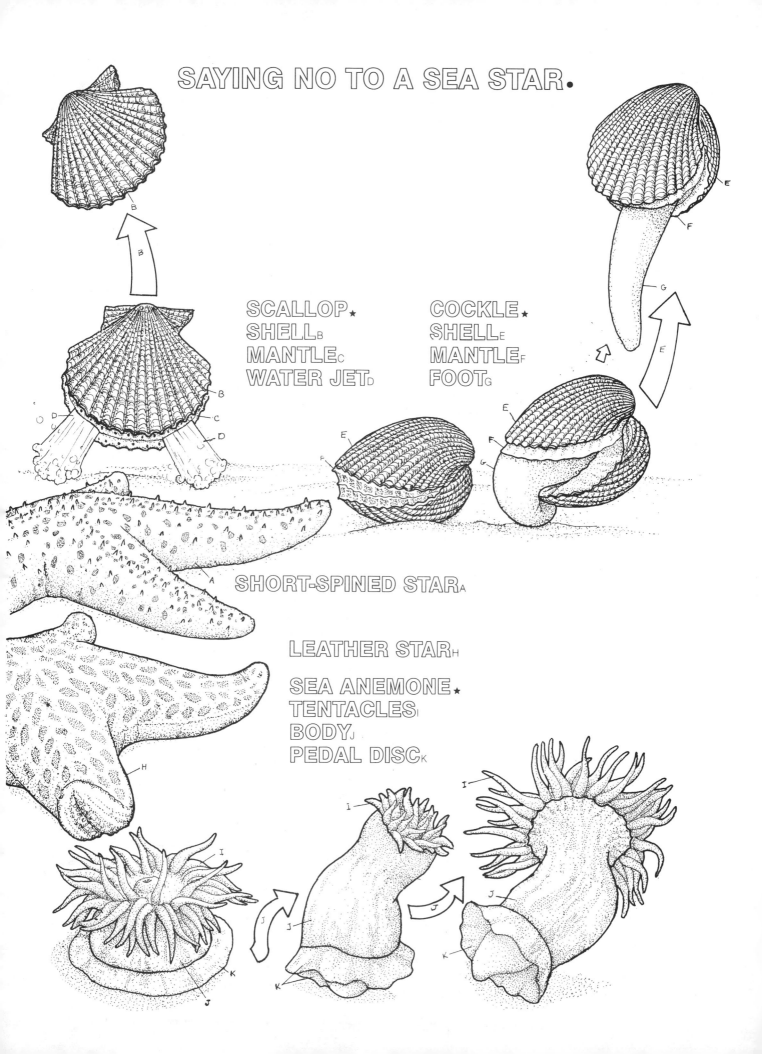

SAYING NO TO A SEA STAR.

SCALLOP ★
SHELL B
MANTLE C
WATER JET D

COCKLE ★
SHELL E
MANTLE F
FOOT G

SHORT-SPINED STAR A

LEATHER STAR H

SEA ANEMONE ★
TENTACLES I
BODY J
PEDAL DISC K

Any consideration of defensive behavior in marine animals must include the cephalopod molluscs with their incredible ability to change color almost instantaneously. Such behavior is employed offensively in capturing prey and in mating squabbles, and defensively, as we will see in this plate in which three instances of escape behavior, or "cephalopod magic" are introduced.

Color the upper row of illustrations, showing defense responses of the squid to the preying dolphin. Save red for the ocelli in the bottom drawing. In the squid the funnel and ink discharge receive the same dark color (gray or black). Color the ink discharge with a ragged edge to suggest its ghostlike shape. Note that the fleeing squid has blanched (it is not colored) and has moved into the background.

The pelagic *squid* does not have a substratum into which it can disappear like the bottom-dwelling octopus. Despite the *squid's* swiftness, there are pelagic predators, such as *dolphin* and tuna, that can overtake and devour them. The *squid,* like the octopus and cuttlefish, has an *ink* sac located near the end of its digestive tract (not shown). This sac empties *ink* into the rectum; it passes out the anus into the mantle cavity. The cephalopod, when pursued, can discharge the contents of this *ink* sac at will, and direct a blob of mucus-bound *ink* into the water through the *funnel.* Simultaneously, the *squid* will darken its entire body by expansion of chromatophores (see Plate 54). Immediately after releasing the *ink,* the *squid* pales markedly by contracting the chromatophores. This sequence of events presents the pursuer with a sudden change in the color of its prey followed by the appearance of a dark, somewhat squidlike mass of unknown origin. The resulting confusion often causes the predator to hesitate, giving the *squid* an opportunity to swim away.

Color the middle illustration of the defense response of an octopus to a grouper. The octopus in dymantic display (F) is left blank except for dark rings around its eyes and a dark margin along its arms.

Octopus species are not only able to change color to match their surroundings, but, with the aid of special muscles, can also alter their skin texture to more closely match the texture of an irregular background. This cryptic (hiding) ability is extremely effective but is not always utilized.

The common *octopus* is a secretive animal which skulks about the rocky bottom, blending into one background after another. However, if confronted by a moving object larger than itself, it often will go into a *dymantic display* (dymantic: threatening). The animal flattens itself against the bottom and changes from a dark color to uniform paleness over its body with the exception of the area around the eyes and along the margin of the interbrachial web (which joins the upper portion of the *octopus'* arms). This response presents the potential predator with a pair of greatly accentuated eyes against a pale background of flesh that exaggerates the *octopus'* size considerably. This image is enough to cause a predator, like the bottom-feeding *grouper,* to pause in confusion, during which time the *octopus* may give a blast of its ink sac and disappear into a crevice!

Color the bottom illustration of the defense response of the *Tremoctopus* to the tuna. *Tremoctopus* receives a different color from its two autotomizable arms. The narrow band encircling each ocellus (J) is left uncolored. The ocelli should be colored red.

The small (20 cm) *Tremoctopus* is prepared to go to great lengths to avoid capture! This pelagic-dwelling octopus can shed, or autotomize (self-cut), its dorsal pair of *arms* to avoid capture. These elongated (15 cm) *autotomizable arms* are much flattened and are kept rolled up when *Tremoctopus* is not disturbed. When threatened, the brightly colored *arms* are unfurled to reveal bright red *ocelli,* or eyespots, outlined in white, lined up along the *arm's* length. Midway between each pair of *ocelli* is a zone of weakness or break point *(autotomy plane)*. This area contains a special muscular arrangement that will allow *Tremoctopus* to release these flattened, *ocellus*-bearing segments sequentially into the water. Once detached, the segments expand in size, probably by the relaxation of partially contracted muscles. In the lower drawing a menacing *tuna* is shown momentarily stunned by the bizarre display of staring, expanding *arm* segments, during which time the octopus escapes. Later *Tremoctopus* will form new segments at the base of these specialized *arms* to replace the ones cast off.

CEPHALOPOD MAGIC.

DOLPHIN_A

SQUID_B
FUNNEL_C ★
INK DISCHARGE_C' ★

GROUPER_D

OCTOPUS_E
DYMANTIC DISPLAY_F ●/÷

TUNA_G

TREMOCTOPUS_H
AUTOTOMIZABLE ARMS_I
OCELLUS_J
AUTOTOMY PLANE_K

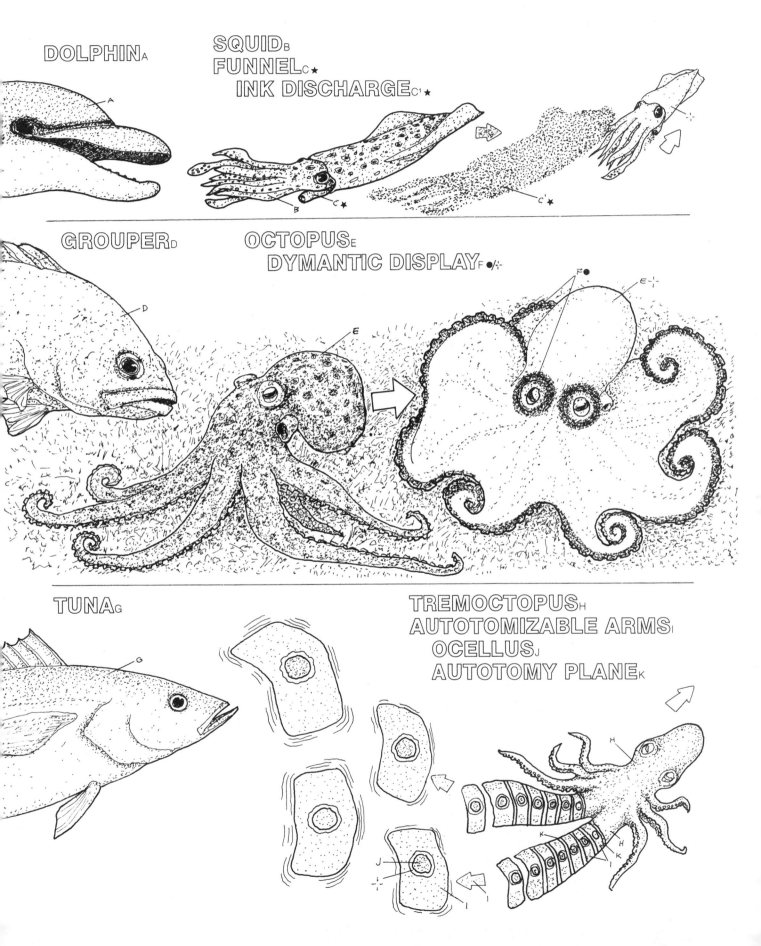

88

FEEDING: MODES OF FILTER FEEDING

The water in many near-shore marine environments is a huge, nutrient-rich "soup." The zooplankton and phytoplankton are suspended throughout the water column. Also small particles of plant and animal remains (detritus) that have been worn down or broken up by wave action are present. Many shallow-water marine animals have evolved ways to reap this floating harvest. Collectively, these animals are called filter feeders because they filter food from large volumes of sea water. In this plate, two members of this interesting group are introduced.

In order to remove suspended food material, a filter feeder must either generate a water current over a collecting device or live where the water moves over the filter. Many animals have cilia that beat rapidly to create a feeding current. These animals also have a filter, covered with a sticky mucus, to trap the suspended food. This type of filter feeding is called ciliary-mucus feeding. The ciliary-mucus combination has been worked to perfection by the bivalve molluscs, sea squirts, and many of the filter-feeding polychaete worms (see Plate 22).

One polychaete that departs from this pattern is the odd-looking but efficient feeder known as *Chaetopterus variopedatus.*

Color the tube (A) first and note that the tube is cut in a longitudinal section and only its edge is colored. The interior of the tube receives the color chosen for sea water (B), which also covers the openings of the tubes. Reserve six light colors for the worm itself. Note that the mucous net (I) is located on the front third of the body and overlaps the body, parapodia, cup, and ciliated groove.

Chaetopterus lives in a U-shaped parchment *tube.* Instead of employing cilia, it generates its feeding current with three pairs of fused *fanlike parapodia* located in the middle of its body. These three parapodial fans beat rhythmically and draw food-bearing water through the *tube.*

The specialized *parapodia* on the twelfth body segment are *winglike* and have many mucus-producing glands, which are constantly employed in the production of a *mucous net.* The *"wings"* of these *parapodia* arch upward against the *tube* and secrete a film of mucus that is carried rearward by the current and forms a *net.* The end of the *mucous net* is caught and held by a ciliated *cup* located in front of the *fanlike parapodia.* This *mucous net* strains suspended plankton and detritus

from the *water* as it is pumped through the *tube.* The end of the food-laden *net* is rolled into a *ball* by the ciliated *cup* until the *ball* is about three millimeters in diameter. Then the *winglike parapodia* release the front of the *net* and the *cup* rolls it all up and deposits the entire *food ball* in a special *ciliated groove.* The celia in the groove move the *ball* forward to the *mouth,* where it is swallowed. Once completed, the whole process begins anew with the secretion of another *mucous net.*

Color the sand dollar. The oral surface, mouth and food groove are to be colored on both the large sand dollar and the square diagram to the left. Then proceed to color spines and tube feet. Finally, color the sand dollars in the sand dollar bed.

The Pacific sand dollar goes one step further than *Chaetopterus.* It uses local water movement, like tidal currents, as its feeding current. In sandy areas with a steady *water current,* the sand dollar burrows partway into the sediment and then hoists itself up on edge so that its body projects above the surface. Directing its flattened *oral surface* into the *current,* the sand dollar catches small particles in weak feeding currents created by cilia (not shown) at the base of the *spines.* These particles are enveloped in mucus and transported to the *food grooves* that lead to the *mouth.* Thus, it was once thought this organism was strictly a ciliary-mucus-feeder.

However, when Dr. Patricia Timko took a closer look at feeding sand dollars, she found food particles in the gut much too large to have been transported by the weak ciliary currents. By close observation she determined that the sand dollar could and did capture large food particles like diatom chains (see Plate 12), and even more active prey like barnacle larvae. In fact, the Pacific sand dollar appears to feed nonselectively on whatever is suspended in the water floating above it, including phytoplankton, zooplankton, and detritus. When an active, suspended animal like a crustacean larva is carried onto the oral surface by the moving *water current,* the small (4 mm) *spines* fold around it to form a cone-shaped trap. The prey is then passed along by the *spines* and *tube feet* to the *food groove.* Once in the *food groove,* the prey is passed along in a coordinated, rowing motion by the stubby *tube feet* that line its side. The prey is delivered to the *mouth* where it is thoroughly chewed by the jaws, and then swallowed.

MODES OF FILTER FEEDING.

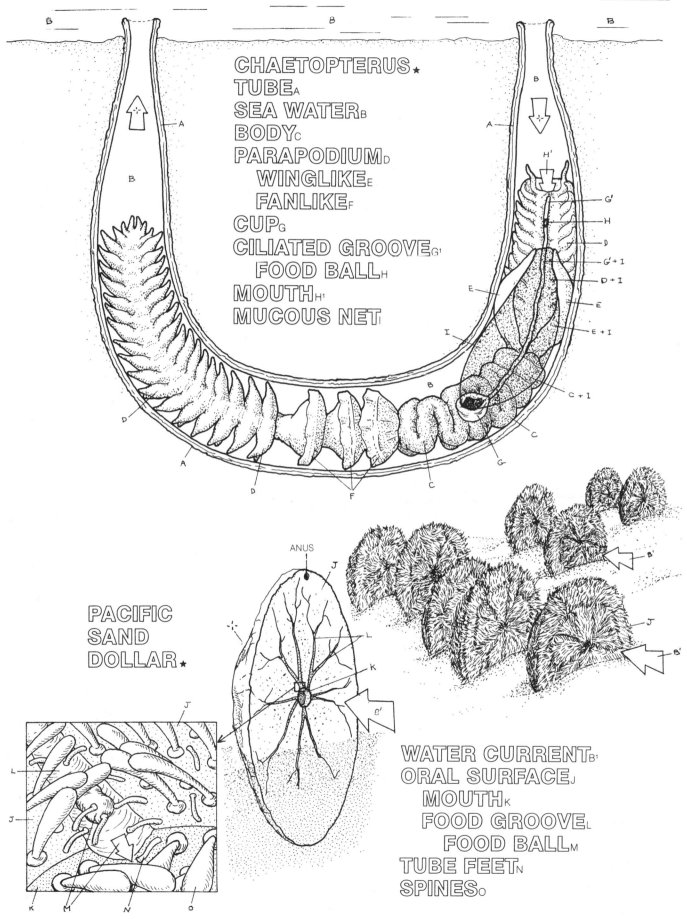

CHAETOPTERUS.★
TUBE A
SEA WATER B
BODY C
PARAPODIUM D
WINGLIKE E
FANLIKE F
CUP G
CILIATED GROOVE G'
FOOD BALL H
MOUTH H'
MUCOUS NET I

PACIFIC
SAND
DOLLAR ★

ANUS

WATER CURRENT B'
ORAL SURFACE J
MOUTH K
FOOD GROOVE L
FOOD BALL M
TUBE FEET N
SPINES O

89

FEEDING: MOLLUSCAN RADULA

In the phylum Mollusca is found one of the most remarkable structures of the animal kingdom—the radula. A radula is basically a flexible file used as a feeding implement in all molluscan classes except the primarily filter-feeding bivalves (Plate 23). The radula demonstrates its greatest functional diversity in the gastropod molluscs (Plates 24 and 25). Members of this class use the radula to brush over the substratum, to grasp and bite, to grasp and suck, to tear up flesh, to bore through shells, and even to harpoon prey (Plate 91).

Color the generalized area of the radula teeth (C) in the ring-top shell snail (upper left) common in California kelp beds. Next color the parts of the enlarged radula area in the central illustration. Note that separate sets of antagonistic (working in opposition) muscles are given the color of the structure that they operate.

In its most general form, the radula consists of a tough *membranous belt* on which are mounted successive transverse rows of sharp *radula teeth*. The *teeth* are formed of a complex of protein and minerals called "chitin" which can be very hard. This toothed *belt* is part of a complex accessory feeding apparatus located just inside the snail's *mouth*. Because the *radula teeth* are so sharp, the radula is confined to a *radula sac* when not in use, so as not to injure the snail's *mouth*. When the snail wishes to feed, the tooth-bearing *membranous belt* is pulled from the protective *radula sac* and the radula is placed against the substratum. The whole feeding apparatus is manipulated by special muscles and cartilage.

As can be seen in the drawing, the toothed *membranous belt* rides over the *odontophore cartilage*. A *protractor muscle* attached to the lower end of the *belt* pulls it down and around the *odontophore cartilage;* the *radula teeth* are facing forward and fold down and slide back over the substratum easily. Once the *belt* is fully protracted (pulled around as far as it will go), then the *membranous belt* is retracted, that is, pulled back around the *odontophore cartilage* by the *retractor muscle* attached to its upper end. This is the effective stroke of the radula (as shown by the directional arrows). As the *radula teeth* come in contact with the substratum they are pulled erect by friction and their sharp cutting edges rasp the surface, tearing pieces free. The pieces are carried into the *mouth* and swallowed. When the snail is finished feeding, the *retractor muscle* attached to the *odon-*

tophore cartilage is contracted, returning the radula to its *radula sac.*

The continued antagonistic action of the *protractor* and *retractor muscles* attached to the *membranous belt* produces a to-and-fro movement that allows the radula to be used like a flexible file on the surface of the substratum. Possession of this marvelous tool allows gastropods to be most effective feeders, as anyone who has snails in their garden can readily testify.

With feeding, the *radula teeth* become eroded. New rows of *teeth* are formed in the *radula sac* throughout the animal's life, and the *membranous belt* grows forward at the rate of several transverse rows of *radula teeth* per day. The worn, anterior *teeth* continuously fall off and the anterior *membranous belt* is resorbed (assimilated) by the snail.

Color only the lower, darkened transverse row of radula teeth in the box at left and the middle row in the box at right. Below each boxed view is the animal possessing the teeth, as well as the food (at left) being chewed or the shell (at right) being drilled.

The size of the radula and the shape and number of *radula teeth* vary considerably among different gastropods; and this is, in part, related to their feeding habits.

The radula of the *abalone* is an example of the feeding tool of a plant eater. The *abalone's* radula has many *radula teeth* per transverse row. The *central* and *lateral teeth* are hooked and sharp, and are used to cut through the abalone's algal (kelp) food. The many small *marginal teeth* collectively act as a broom to sweep the dislodged algal pieces into the snail's mouth.

In contrast, the carnivorous *oyster drill* uses its radula to drill through (rasp) the shells of oysters and other molluscs. There are only three *radula teeth* per transverse row. The *central tooth* is very large and has three long, sharp cusps (cutting edges). These are used in drilling the shell, which is first softened by the acid secretion of a special gland located on the snail's foot. The two *lateral teeth* are hooked and protrude at right angles to the *central tooth*. These operate as "meat hooks" to grasp and pull in the pieces of tissue from the body of the prey once the *hole* is gouged through the shell. The *oyster drill* takes approximately eight hours to drill through a shell two millimeters thick.

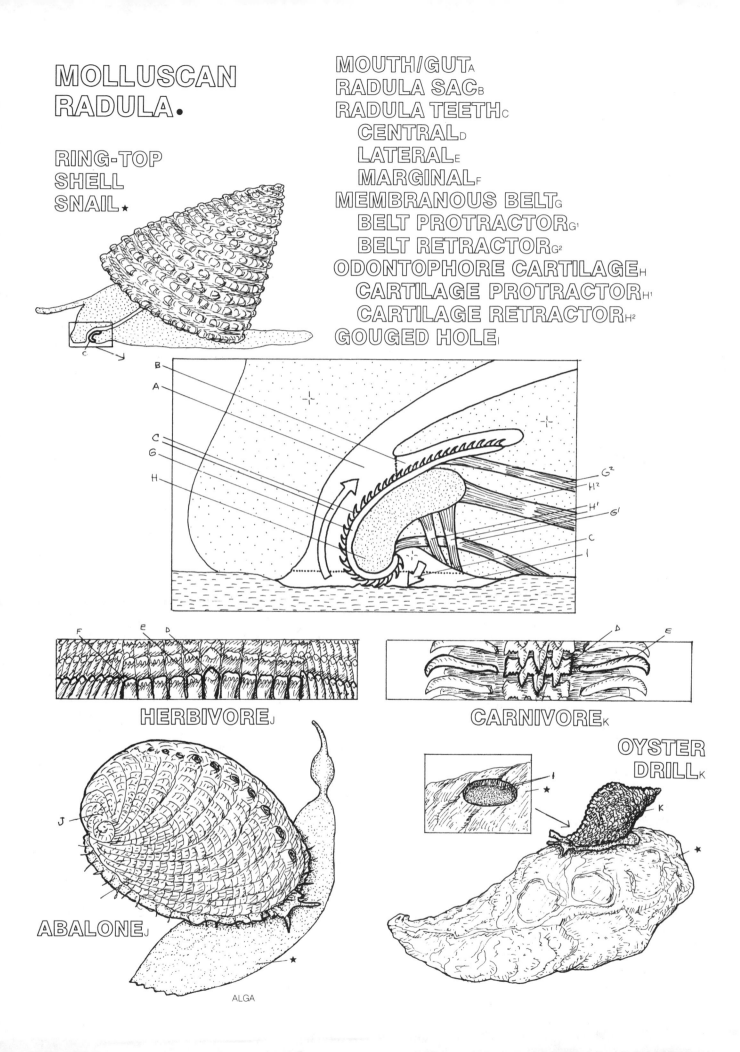

MOLLUSCAN RADULA.

RING-TOP SHELL SNAIL ★

MOUTH/GUT A
RADULA SAC B
RADULA TEETH C
  CENTRAL D
  LATERAL E
  MARGINAL F
MEMBRANOUS BELT G
  BELT PROTRACTOR G¹
  BELT RETRACTOR G²
ODONTOPHORE CARTILAGE H
  CARTILAGE PROTRACTOR H¹
  CARTILAGE RETRACTOR H²
GOUGED HOLE I

HERBIVORE J

CARNIVORE K

ABALONE J

ALGA

OYSTER DRILL K

The large marine plants present a bountiful source of food for marine invertebrate plant eaters, or herbivores. In this plate, four invertebrate herbivores from the west coast of the United States are introduced.

Start in the lower left corner and color the sea urchins and the giant kelp plant. Next move to the upper right corner and color the enlarged piece of drift kelp and the beach hoppers. Note that the drift kelp was detached from the giant kelp (B) plant below (follow arrows). Proceed down the right side of the page as each herbivore is treated in the text. Note that the chiton's shell plates and the coralline algae receive the same color. If you wish, leave the thinner lines on the shell plates blank, as they are white in life.

Sea urchins are a common sight in California *kelp* beds. They may be seen on the surface of rocks or in crevices where they wait for algae to drift by. They grab the *kelp* with their suckered tube feet and devour it with their five-jawed Aristotle's lantern. However, if too many *sea urchins* collect in an area, drift algae is insufficient to support them. Then the hungry herbivores go on the march looking for food.

The most vulnerable point on the *giant kelp* plant is the stipe just above the holdfast (see Plate 14). Hordes of marauding *sea urchins* enter *kelp* beds, clamber up the holdfast and eat through the stipes, sometimes releasing a 12-meter-long plant from its contact with the bottom. The huge *kelp* plant will wash either out to sea or up on the beach. *Sea urchin* hordes or ''fronts'' have stripped acres of *kelp* forest habitat in this manner. Herbivore population explosions are kept in check in areas where sea otters forage, as will be seen in Plate 96.

The drift algae that wash up on the beach feed huge populations of *beach hoppers.* These herbivores are large (to 4 cm) amphipod crustaceans, which burrow in the sand high on the beach by day and emerge to feed during nocturnal low tides. These animals are equipped with strong-biting mouth parts (not shown) and devour copious amounts of drift algae during even a single tide. They appear to prefer the *kelp* species, perhaps because they are more tender and easier to chew. They also go after the freshest algae on the beach. They play a significant role in breaking up large seaweeds into much smaller pieces that can then be used by other marine animals.

Another *kelp* lover is the small (15 mm) *limpet,* that both eats and makes its home on the *feather-boa kelp.* This *kelp* grows in the low rocky intertidal and subtidal zones. It has many straplike stipes that may reach a length of six meters. The *limpet* feeds along the midrib of the *kelp,* leaving a depression or *grazed area* where it has eaten. These *grazed areas* are excavated by the *limpet* with its specialized radula. It lives in one of these depressions, called a home scar.

One of the most strikingly colored intertidal animals of the west coast of the United States is the lined chiton. This small (5 cm) mollusc occurs in a number of different vivid color variations, including deep violet, pink, and creamy brown. Its eight *shell plates* are always highlighted by white lines. The lined chiton is one of the few animals that seems to feed on tough *coralline algae.* These red algae sequester calcium carbonate (lime) from sea water and use it to impregnate their cell walls, giving the *algae* a crusty, plasterlike texture that discourages most herbivores. The lined chiton appears to feed on the *coralline algae* and perhaps even utilizes the *algae's* pigments in its shell. This has not been shown conclusively, but the colors in the chiton's *shell plates* often match the *coralline algae* with which it is associated.

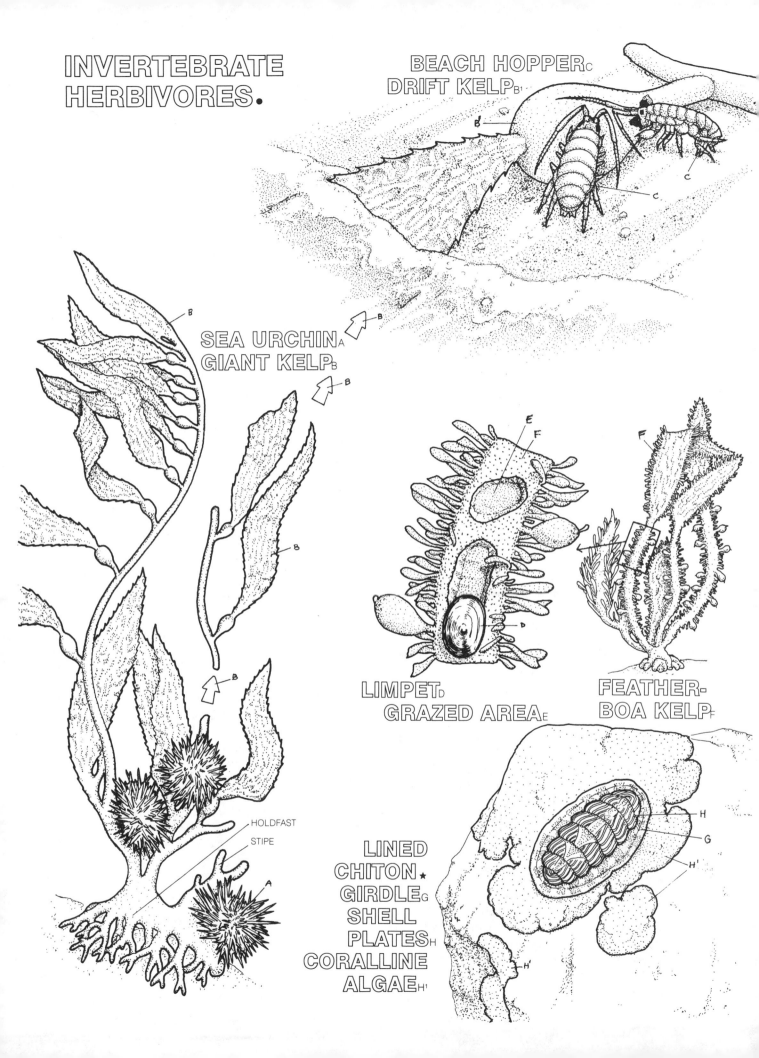

INVERTEBRATE HERBIVORES.

BEACH HOPPERc
DRIFT KELPв₁

SEA URCHINᴀ
GIANT KELPв

LIMPETᴅ
GRAZED AREAᴇ

FEATHER-
BOA KELPꜰ

HOLDFAST
STIPE

LINED
CHITON ★
GIRDLEɢ
SHELL
PLATESʜ
CORALLINE
ALGAEʜ₁

91

FEEDING: INVERTEBRATE PREDATORS

There is a tendency to classify all invertebrates as sluggish, plodding, unexciting animals. This fallacy can be quickly dispelled with many examples, such as the predatory invertebrates. The techniques of these animals compare equally with or excel many of the predaceous methods of their vertebrate counterparts.

Begin with the cone snail and color each predator separately as it is treated in the text. The victims are to be left uncolored in the five examples shown.

The cone snails employ "harpoons" in dispatching their prey. These snails (various species, 2–25 cm) are most abundant in tropical and subtropical Pacific and Atlantic shallow-water habitats. Here they feed on a variety of prey, ranging from polychaete worms to small fish. The "harpoon" is actually a single, specialized *radula tooth* that is equipped with a spearlike barb at the tip. The *tooth* has a groove along its length through which the snail injects a quick-acting poison from a gland located adjacent to the radula sac (not shown). When a prey animal is located, the *proboscis* is moved in close and the *radula tooth* "harpoon" is fired into the prey by contraction of a large muscular bulb (not shown). The prey animal is quickly subdued by the poison. The *radula tooth* remains attached in fish-eating species and is "reeled" in by retraction of the *proboscis*. The fish is swallowed whole by the *mouth,* which is capable of incredible distension. Those species that specialize in capturing fish have a poison that is potentially fatal to humans. Thus, cone snails and their beautiful *shells* are best appreciated at a distance.

Another aggressive invertebrate is the crustacean known as the *mantis shrimp.* They are so named because their stalked eyes and large, elevated *claws* give them a striking resemblance to the insect known as the praying mantis. There are approximately 300 species of *mantis shrimp,* ranging in size from five to thirty centimeters. They are found in burrows on soft bottoms or in rock or coral crevices. Most species feed on soft-bodied invertebrates like snails or shrimp, but some catch fish. *Mantis shrimp* usually lie in wait at the front of their burrows until prey comes near; then they quickly swim out and slash the prey with their large *claws.* The last segment of the *claw* is fitted either with sharp spines or a knifelike blade. The "blade" folds backward into a depression in the *claw* segment behind it. *Mantis shrimp* have nasty tempers and have earned the name "split thumb" from fishermen trying to remove them from their nets.

A more subtle, but no less lethal, predator is the large (15 cm in diameter) beaded *sea anemone* shown here with a bat star in the embrace of its *tentacles.* This predator waits for prey to venture near, and then its tentacles discharge hundreds or thousands of nematocysts into the victim. It can quickly subdue even large prey such as sea stars, although they are a rare catch for anemones.

Stretched across the page with a lobster in its grasp, is the *arm* of a large *octopus.* Crabs and lobsters are common prey for the eight-armed *octopus.* When captured, the victim is pinned down with the sucker-bearing *arms* and deftly bitten by the birdlike beak; then poison is pumped into the prey at the site of the bite by a special hydrostatic organ. The *octopus* floods the victim with digestive enzymes, sucks in the partially digested tissues, and discards the shell as an empty husk. For humans seeking to catch octopuses, the bite of this animal is of much greater concern than encirclement by tentacles.

The predatory habits of the *crown-of-thorns* sea *star* recently caused quite a stir in several areas of the South Pacific. This sea *star* specializes in eating coral, and prefers the kind which are principal reef-building corals. Like many sea stars, the *crown-of-thorns* can evert its stomach (not shown) onto its prey such as the brain coral shown here, and secrete digestive enzymes onto the fleshy parts. The prey is digested outside the *star's* body and the dissolved parts are absorbed by the everted stomach. There appeared to be a "population explosion" of sea *stars* in the areas just mentioned, and there was fear that whole reefs would be destroyed as great numbers of sea stars devoured the coral organisms. However, cores taken from Australian reefs suggest fluctuations in numbers of *crown-of-thorns* sea *stars* have occurred for thousands of years and, thus, the recent "population explosion" may fall within its normal range of fluctuations.

INVERTEBRATE PREDATORS.

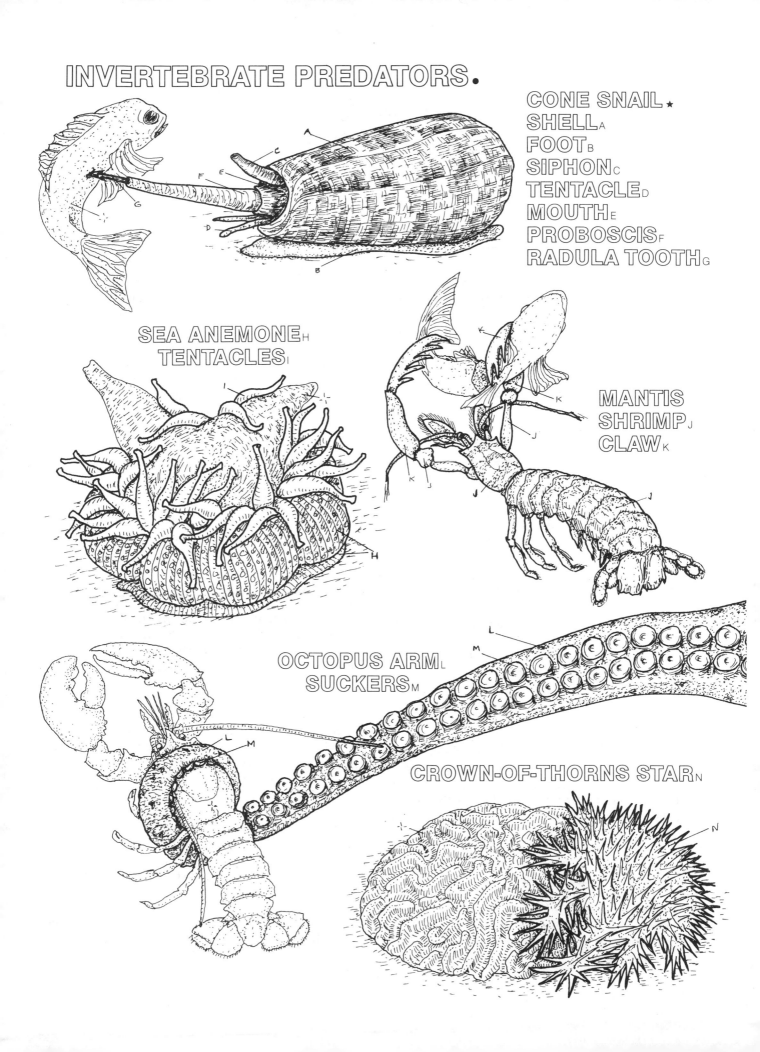

CONE SNAIL. ★
SHELL A
FOOT B
SIPHON C
TENTACLE D
MOUTH E
PROBOSCIS F
RADULA TOOTH G

SEA ANEMONE H
TENTACLES I

MANTIS
SHRIMP J
CLAW K

OCTOPUS ARM L
SUCKERS M

CROWN-OF-THORNS STAR N

92

FEEDING IN BONY FISHES: ATTACKERS AND AMBUSHERS

Seeing a marine habitat such as a coral reef for the first time, one is invariably entranced by the beauty and variety of fish present (see Plate 38). Upon seeing so many different kinds of fish in such numbers, one wonders how the habitat can support such a wide diversity of fish without severe competition. Part of the answer lies in the variety of ways marine fishes feed. In the next three plates, different kinds of feeding strategies of marine fish will be explored, together with the corresponding range of behavioral and morphological adaptations.

Color each predatory fish separately as it is treated in the text. The prey fish are to be colored gray. The shark on the right side of the page, although an attack-type predator in its own right, is not colored. Note that the frogfish has a lure (B¹) which is a modified dorsal fin spine, and receives the dorsal fin color.

The most well-known feeding strategy is that of the pursuing carnivore that "runs down" its prey (usually other fish or active invertebrates), attacks, and devours them. Such fish are usually highly visible. One example is the great barracuda which is found worldwide in warm and temperate waters except the eastern Pacific and Mediterranean. This fish is very long (up to 3 meters), streamlined, muscular, and possesses a large falcate (scythe-like) *caudal fin* which is used for quick acceleration and rapid swimming over short distances. The barracuda is a crafty, curious fish and often approaches divers, but always remains just out of reach. Any quick movement toward it results in its swift departure. The barracuda has long *jaws* with long sharp teeth (many of which curve backward) for seizing and holding prey fishes. When a fish is too large, the barracuda may cut it in two and return to gulp the pieces.

This slashing attack is also characteristic of the bluefish schools of the western Atlantic. So swift and menacing is this medium-sized predator that it is accused of tearing into schools of terrified bait fish and killing far more than it can eat.

Other pursuing predators are the jacks, which are capable of amazing acceleration and speed. They have been observed stealing meals out of the mouths of

sharks feeding on fish tossed from boats. Pictured here is the greater amberjack, known in all tropical and subtropical seas. The amberjack usually has a brass-colored stripe along the body at the level of the eye; it is olive to brownish above the stripe and silvery white below. This fish can get quite large: there are hook-and-line records of 1.4 meters and over 63.5 kilograms (140 lbs). The amberjack exemplifies the body form of a fast-swimming pelagic fish with its torpedo-shaped *body* and large lunate *caudal fin.* Jacks tend to swim in groups and range over large areas. Though not residents of coral reefs, they will swim over reefs and feed on local fishes.

A more leisurely predation strategy is to "sit and wait." The odd-looking trumpetfish (discussed in Plate 38) employs this tactic, as does the lizardfish. Appropriately named, this slender, lizard-shaped fish has a big *mouth* full of reptile-like *teeth.* The fish illustrated here (the rockspear lizardfish) occurs on both sides of the Atlantic and reaches a length of 33 centimeters. It is basically silver-white in color with mottled reds and browns, and blends in with the mud and sand bottoms that it frequents. Lizardfish sit quietly on the ocean floor, sometimes partially buried, and wait for small fish to swim overhead, at which time they dart upward and seize their prey in their well-equipped *mouths.*

A slight variation on this ambush technique involves a trick that has been named "aggressive mimicry" by some scientists and is seen in a strange group called the frogfish. These fishes, represented here by the splitlure frogfish, have a modified first *dorsal fin* spine on their snouts. This *lure,* or illicium, is wiggled about enticingly to attract unsuspecting fish. The frogfish remains immobile on its peculiar stumpy *pectoral fins,* looking like a sponge on a coralline-algae-encrusted rock. When a fish comes in to grab the *lure,* the frogfish sucks the prey into its cavernous *mouth* by rapidly expanding its mouth cavity. Recent studies of related frogfish species have shown that they can expand their mouth cavity to 12 times its normal resting volume within 6–10 milliseconds. This is one of the fastest of all capture mechanisms in the animal kingdom.

ATTACKERS & AMBUSHERS.

BODY A
FINS ★
   DORSAL B
   CAUDAL C
   ANAL D
   PELVIC E
   PECTORAL F
EYE G

JAWS/MOUTH H
PREY I ★

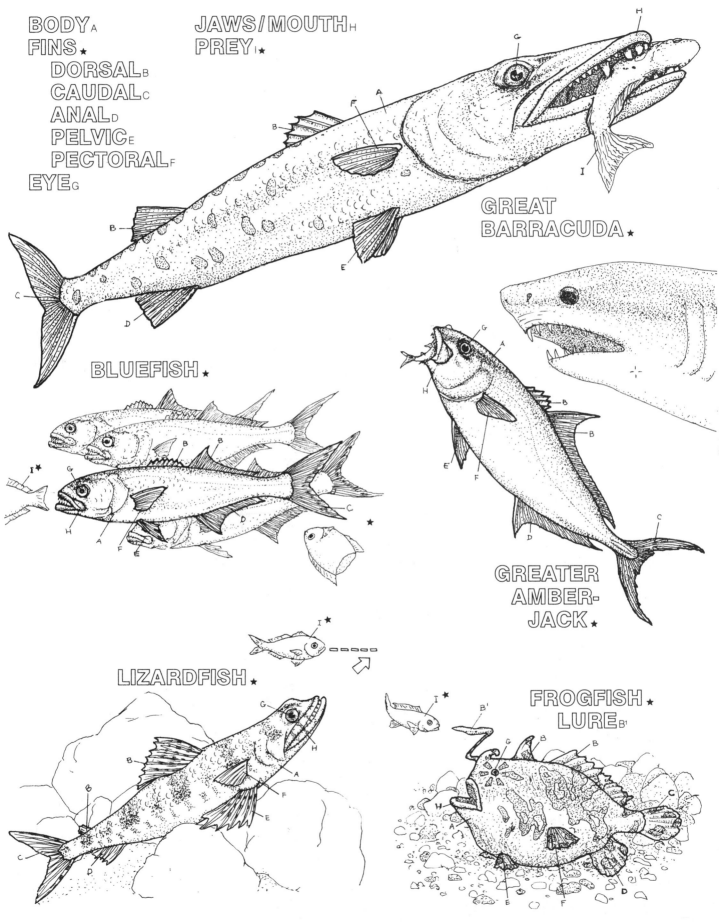

GREAT BARRACUDA ★

BLUEFISH ★

GREATER AMBER- JACK ★

LIZARDFISH ★

FROGFISH ★
LURE B'

93

FEEDING IN BONY FISHES: PICKERS, PROBERS, AND SUCKERS

In this and the next plate, fishes that feed on a variety of prey besides other fish will be considered. Many fishes ("generalists") feed somewhat opportunistically, on whatever is available, and many others will eat a small variety of prey species. Placing fishes into feeding categories is sometimes arbitrary, and there may be considerable overlap between categories.

Color each fish as it is treated in the text. Begin by coloring the California sheephead and its prey (upper drawing). Read the text relating to the pipefish and color the drawing of the entire animal, then proceed to the blown-up inset showing its mouth and prey, to the left. Follow the same procedure for the filefish and the butterflyfish. Finally, color all views of the triggerfish and its sea urchin prey.

The California sheephead is a familiar sight in U.S. west coast kelp beds. This large (up to 1 meter) orange and black fish (in the case of the males; the female is solid orange) feeds on invertebrates and is considered a "generalist." Sheepheads have stout *jaws* and large protruding *teeth* for crushing a variety of prey. They are known to feed on sea urchins, sand dollars, mussels, scallops, abalone, lobsters, hermit crabs, octopus, tube-dwelling polychaetes, or any small to moderate-sized marine invertebrate it finds in the kelp habitat.

Prey selection is much more limited for the pipefish. Pipefishes and the closely related seahorses are delicate suctorial feeders. They use their tubelike snouts and small *mouths* to ingest food by a rapid intake of water, as if sucking on a straw. They have prehensile grasping tails that they use to anchor themselves on algae and other substrata. Their *eyes* can move independently, like those of a chameleon, allowing them to scan the water for their small prey from their anchored perch. Pipefishes are poor swimmers; their *caudal fins* are reduced and the *dorsal* and *pectoral fins* provide the main swimming thrust. Pelvic fins are absent. There are approximately 150 species of pipefishes, found mostly in the shallow waters of tropical and subtropical seas; a few are freshwater inhabitants. They are generally small (15

cm), with the largest reaching 50 centimeters.

The reef filefish of the tropical Pacific feeds on small reef invertebrates by probing into crevices with its elongated snout and using its small, sharp, incisor-like teeth to pick out prey. This small (10–30 cm), brightly colored (green with orange spots) fish is often seen among the branches of various species of coral and is known to feed primarily on coral polyps. Filefish are closely related to triggerfish; they have a similarly large first dorsal spine, but lack the triggerfish's locking mechanism (Plate 85). Their skin is rough to the touch and was once utilized as a kind of sandpaper.

The forceps butterflyfish of Pacific coral reefs also picks and probes for food. Sporting the familiar butterflyfish eyespot (Plate 49), this small fish (10–15 cm) has an elongated snout and pointed *jaws* with sharp teeth that are used like a pair of forceps to extract small morsels from tight quarters. Its principal food items are the tentacles of tube-dwelling polychaetes, coral polyps, and tube feet and pedicellariae plucked from between the spines of sea urchins. Like many butterflyfishes, this species raises its stout *dorsal fin* spines as a defense mechanism, presenting an unappetizing mouthful to swallow.

The queen triggerfish is a relatively large fish (up to 26 cm) and a generalist feeding on invertebrates. Instead of the large *jaws* of the sheephead, it has a smaller *mouth* with short, stout *jaws,* each with eight protruding incisorlike *teeth.* It uses this equipment to render the hard portions of its molluscan and crustacean *prey* into small pieces. The queen trigger prefers the long-spined sea urchin, which presents a formidable defensive posture. Triggerfishes have a tough leathery hide made of bony scales that provide a flexible armor against the sea urchin's spines. The queen trigger lifts the urchin by a single spine, carries it up off the substratum and drops it. The urchin usually lands oral side down, but after repeated upheavals the urchin will eventually fall with its vulnerable oral area exposed. The queen trigger then darts in between the shorter oral spines, quickly bites through the soft mouth area, and proceeds to eat the urchin from the inside out.

PICKERS, PROBERS & SUCKERS.

BODY A
FINS ★
    DORSAL B
    CAUDAL C
    ANAL D
    PELVIC E
    PECTORAL F
EYE G
JAWS/MOUTH H
    TEETH H'
PREY ★

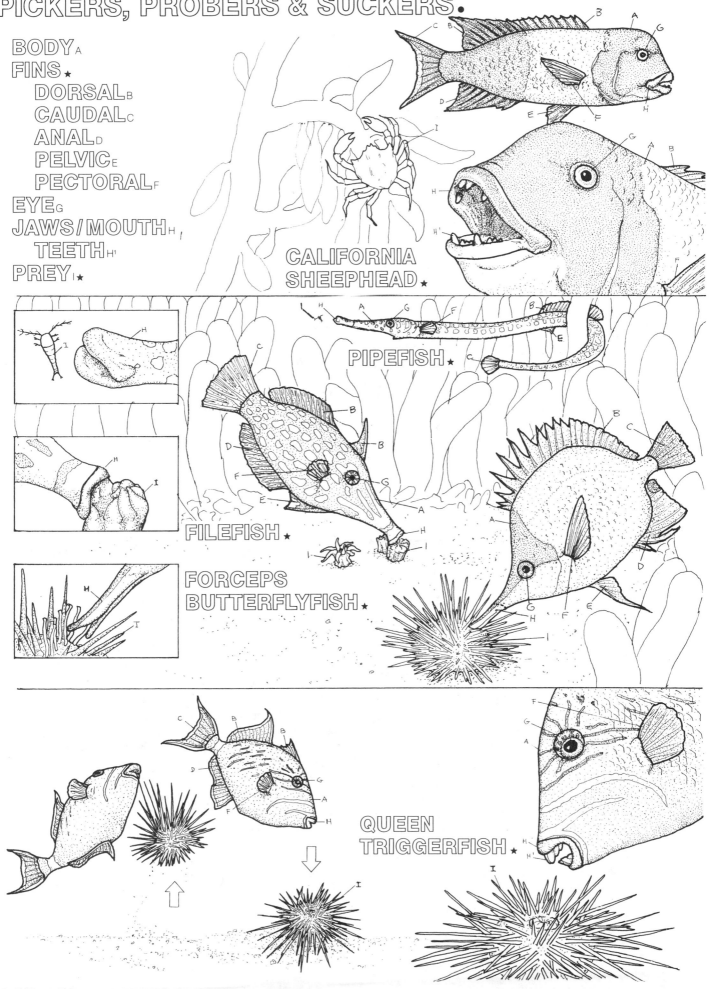

CALIFORNIA SHEEPHEAD ★

PIPEFISH ★

FILEFISH ★

FORCEPS BUTTERFLYFISH ★

QUEEN TRIGGERFISH ★

Some fishes, like the barracuda or the sheephead, attack, chase, and consume whole prey. Other fishes, like the reef filefish and forceps butterflyfish, probe and pick away at parts of their prey. In this plate, fish that graze (like the continual browsing of cows or sheep), and those that grub through soft substrata seeking their prey, are considered.

In sequence color each fish together with the diagram of its mouth on the right. The prey receive the same color in all cases. Note the trunkfish lacks pelvic fins. Note that the bat ray's mouth is not visible in the larger view; in the smaller view, it is shown from the ventral surface. Note that the ray lacks caudal and anal fins.

The northern anchovy represents a group of grazers that utilize the vast pasture of the plankton. These small fish (to 22.5 cm) swim through the plankton-rich water with their large *mouths* open and trap the plankton on their gill rakers. They will occasionally go after a larger plankter and snap it up. Anchovy species are found in all the world's oceans, and are important as bait and food fish.

The parrotfishes of Atlantic and Pacific coral reefs are also grazers. These medium- to large-sized fish (25–100 cm) are primarily herbivores, feeding on the low growth of algae found on coral rock. The parrotfish's *teeth* are fused into a stout beak with sharp cutting edges. These are used to scrape the algal growth from coral rock, and sometimes to ingest living coral as well, supposedly for the plant cells contained in the coral polyps (see Plate 76). The ingested algae and coral are ground up by a "pharyngeal mill" consisting of molarlike teeth on the floor and roof of the throat (not shown). This feeding activity turns coral rock into coral sand and, in local areas, can account for substantial sediment increase and coral reef destruction. A single parrotfish may turn a ton of coral reef into sand in a year's time.

Many marine invertebrates live within the soft substrata that form the dominant types of ocean bottom. This infauna includes clams, crustaceans, polychaetes, and many others. Frequently, the predator's problem is to discover them. Fishes that feed on these buried, bottom-living forms are often called grubbers, suggesting the chore of sorting through sand or mud to find food.

The trunkfishes are sophisticated grubbers. Instead of mucking about in the substratum they swim almost vertically above it and direct jets of water from their *mouths* downward onto the surface of the substratum. This removes the upper sediment layer and exposes the buried prey, which are then consumed. Trunkfish feed primarily on small crustaceans and polychaete worms.

The bat ray is a most determined grubber. This large, 1-meter wide, 95 kg (210 lb) elasmobranch fish uses its winglike *pectoral fins* to fan away the substratum and expose burrowed prey. It feeds primarily on bivalves and crustaceans, which are readily crushed by its stout *jaws* and flat pavementlike *teeth*. The bat ray has been reported to use the undersurface of its wing like a giant suction cup to suck buried prey out of their burrows.

A bit more grubbing finesse is demonstrated by the familiar goatfish of tropical waters. These medium-sized fish (25–50 cm in length) possess a pair of chemosensory (chemical-sensing) appendages called *barbels* that hang from their lower *jaws* like chin whiskers. The flexible *barbels* are moved rapidly over the substratum or through it, detecting buried *prey* (small crustaceans and worms) which are quickly excavated and devoured by the small, downward-facing *mouth*. There are approximately 50 species of goatfish, and most of them are found inshore in shallow water. Individual species show specific feeding preferences for sand or mud substrata and also for feeding times (some are nocturnal feeders). Many species are characterized by group feeding in moderate-sized schools of 25 to 50 individuals.

GRAZERS & GRUBBERS.

BODY A
FINS ★
  DORSAL B
  CAUDAL C
  ANAL D
  PELVIC E
  PECTORAL F
EYE G
JAWS/MOUTH H
  TEETH H¹
BARBELS I
PREY J ★

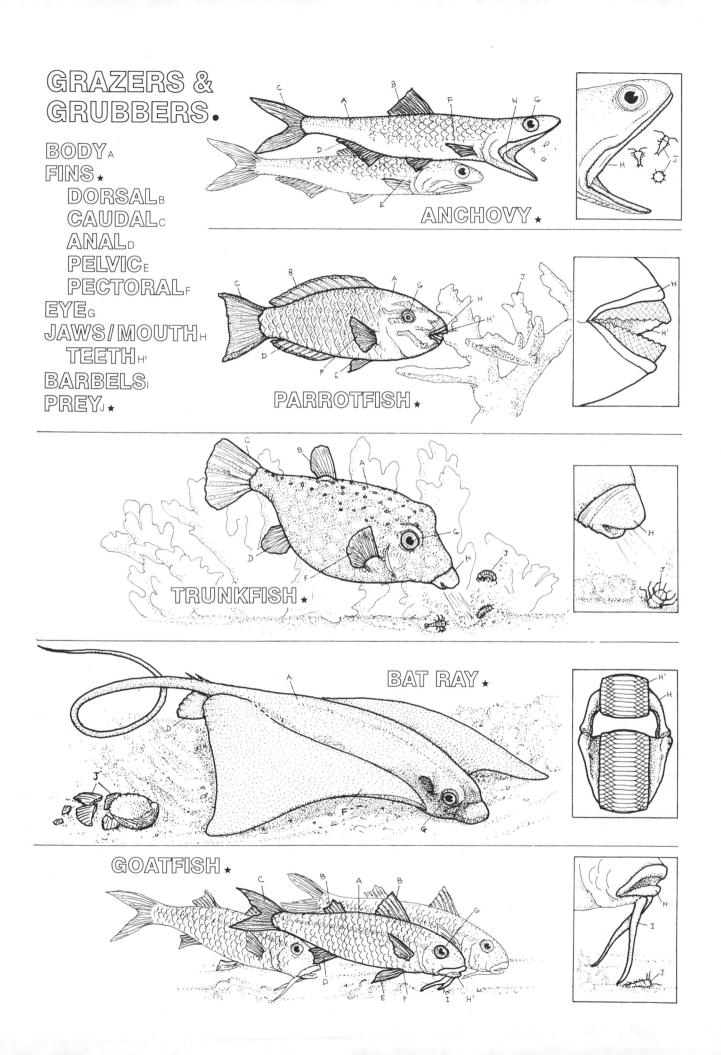

ANCHOVY ★

PARROTFISH ★

TRUNKFISH ★

BAT RAY ★

GOATFISH ★

The marine rocky intertidal zone is a complex and fascinating habitat. Some individual species form distinct zonation patterns that are characteristic of this habitat worldwide (see Plate 2). The success of an individual species in this habitat depends both on its ability to withstand physical stress and on its successful interaction with other intertidal organisms. In the rocky intertidal zone the main interspecific interactions are predation and competition for space. In this plate we will seen how the interaction between a predator, the *sea star* and its preferred prey, the *mussel,* plays a key role in structuring (developing) and maintaining intertidal zonation.

Before beginning to color, read through the entire text. The illustration represents the same habitat with (top) and without (bottom) the sea star. Color the sea star first and then the mussels in both illustrations. Color the remaining organisms as they are mentioned when you re-read the text. The difference in diversity of organisms (and color) will indicate the importance of the sea star's predatory role.

The California sea *mussel* occurs only in areas exposed to wave action; along the Pacific coast this *mussel* forms a conspicuous band in the middle intertidal zone. The upper limit of this solid band of *mussels* seems to be determined by tidal exposure. The *mussel* can tolerate exposure to air for only a limited period; the *mussel* is absent from intertidal areas uncovered for periods beyond its exposure tolerance. The lower limit of the band of *mussels* is determined by the predatory behavior of the *sea star.* The *sea star* can capture and eat almost all the invertebrates of the middle intertidal, but it prefers the *mussel* over all other potential prey. When the tide is in, the *sea star* moves upward and preys on the lower *mussels.* However, the *sea star* must successfully find and eat a *mussel* and return to a lower position before the tide goes out; it cannot tolerate the length of exposure to air encountered at the level of the *mussel* band. Because of this exposure limitation, those *mussels* that occur above the *sea star's* reach are safe from this dominant predator, and form the *mussel* band.

In the middle intertidal zone, below the *mussel* band, a variety of other organisms that are tolerant of wave action occur, including large algae like the *feather-boa kelp* and *sea palms.* Large patches of the *stalked barnacle* and the solitary giant *green sea anemones,* as well as herbivorous molluscs like the *limpets* and *chitons* are found here.

The predatory interaction between the *sea star* and the *mussel* was studied for over ten years by Dr. Robert Paine of the University of Washington. Dr. Paine selected wave-exposed, rocky intertidal habitats with distinct *mussel* bands along the Washington coast. In certain areas he removed all the *sea stars,* leaving adjacent areas unaltered in order to compare the growth of the *mussel* band with and without *sea star* predation.

In the areas where the *sea star* had been removed, the *mussel* band began to extend itself downward. This increase in its width was caused by the growth of the *mussels* already present, and by the settling of new *mussel* larvae from the plankton (see Plates 60 and 63). The *mussel* larvae settled on the shells of the adults and among the filaments of *red algae* growing near the *mussel* band. In the absence of *sea star* predation, the population of *mussels* grew and extended its range downward, eventually carpeting the rocky substratum with a layer of *mussels.* As the *mussels* colonized the lower intertidal zone, attached organisms like *stalked barnacles* and *sea anemones* were encircled and gradually squeezed to death. Large algae were overgrown and smothered, and herbivorous invertebrates were deprived of food and space for attachment.

In the adjacent areas where no *sea stars* had been removed, the width of the *mussel* band remained unchanged through the experimental period. When *sea stars* were allowed to re-invade the experimental *sea star* removal areas, their predation began to push the lower edge of the *mussel* band upward toward its original position.

Thus the *sea star's* predation on *mussels* restricts their extension downward into lower zones. This prevents the *mussel* from monopolizing the available space in the middle intertidal zone, and assures space for the attachment and growth of other intertidal organisms.

THE SEA STAR/MUSSEL INTERACTION.

SEA STAR A

MUSSEL B

   LARVA B¹   GREEN SEA ANEMONE F

FEATHER-BOA KELP C   LIMPET G

SEA PALM D   CHITON H

STALKED BARNACLE E   RED ALGA I

Read the entire text before coloring this plate. The left frame shows the sea otter at work in a kelp bed. The right frame represents an area of former kelp bed habitat.

The *sea otter* is commonly seen in the *kelp* beds along the central California coast. This whiskered, playful marine mammal plays a leading role in a drama that has both biological and historical implications. The *sea otter* once occurred in a continuous arc along the west coast of North America from Baja California to Alaska and the Aleutian Islands. It was hunted relentlessly in the eighteenth and nineteenth centuries by Russian fur traders for its fine, thick fur. The fur hunters eliminated the *otters* from much of the latter's historical geographical range. Along the west coast of the United States, only a small colony survived, unnoticed until the 1930s, near Big Sur, California. This colony of *sea otters* has increased in numbers, and individuals are moving north and south along the central California coast.

Sea otters lack the insulating blubber layer found in most other marine mammals, and therefore they must consume up to 25 percent of their own body weight in food each day to maintain their body heat in the cold Pacific water. Because of this requirement, *sea otters* spend a good deal of their day hunting and eating large invertebrates and occasionally fish. When they enter a new area they first devour the largest, most abundant invertebrate prey. In a *kelp* bed this diet includes principally *abalone* and *sea urchins*. In *kelp* beds where *otters* have been feeding regularly, the only remaining *sea urchins* and *abalone* occur under rocks and in deep crevices well out of the reach of the *otters*. Broken *abalone* shells and empty *urchin* tests (skeletons) litter the bottom and attest to the *sea otter's* hunting skill.

The absence of *sea otters* in *kelp* beds has been shown to have significant biological ramifications. A disturbing process has occurred in southern California, where *kelp* beds once inhabited by *sea otters* have gradually disappeared and not returned. Divers discovered that these barren areas were inhabited by *sea urchins* and patches of encrusting *coralline algae* that are unpalatable to *urchins* (see Plate 13). It was determined that the *urchins* (no longer preyed upon by *sea otters*) had increased in number and had moved through the *kelp* bed in a veritable "*urchin* front," destroying all the plants in their path. No young, growing *kelp* or any other plants escaped these organisms. After the *kelp* bed was eliminated, some *urchins* remained alive, perhaps by their ability to absorb dissolved organic material from nearby sewage outfalls. These starving *urchins* prevented the re-establishment of the *kelp* beds.

It appears that the *sea otter* plays a key role in maintaining the *kelp* forest habitat. By feeding on *sea urchins,* the *sea otter* reduces their population to a point where they have minimal impact on the *kelp* plants. The presence of the *kelp* bed creates a rich near-shore habitat (see Plate 8) populated by many organisms.

Many conservationists believe that as the *sea otter* population spreads along the California coastline it will significantly alter the shallow near-shore environment. Areas now dominated by *sea urchins* will perhaps be opened up for the establishment of *kelp* beds or other rich algal habitats. However, recent studies indicate that increased growth of the *kelp* bed canopy severely reduces the sunlight reaching the benthic (bottom) communities and inhibits their growth. The *kelp* forest attracts certain species of fish but also eliminates habitat for some other, commercially important fish species that prefer open water.

The potential spread of the *sea otter* poses other biological and social problems. Because of their voracious appetites and opportunistic feeding habits, *otters* threaten to reduce, and perhaps eliminate the *abalone* fishery of central California. The *sea otters* will also forage along sandy bottoms and dig for clams. Clammers fear they will eliminate the extensive Pismo clam beds of central California's sandy beaches. Finally, there is the potential of the *sea otters* to eat themselves out of "house and home"; a herd of 2,000 *otters* may consume up to 20 million pounds of invertebrates in a single year.

SEA OTTER A

KELP B
ABALONE C
SEA URCHIN D
TEST D'
CORALLINE ALGAE E

APPENDIX OF
SCIENTIFIC NAMES

22 Cockle clam, *Clinocardium nuttalli*

23 Scallop, *Pecten*
Mussel, *Mytilus edulis*
Cockle, *Clinocardium nuttalli*
Bent-nosed clam, *Macoma nasuta*
Softshell clam, *Mya arenaria*

24 Tulip snail, *Fasciolaria tulipa*
Abalone, *Haliotis*
Moon snail, *Polinices lewisii*
Cowry, *Cypraea*

25 Dorid nudibranch, *Diaulula sandiegenesis*
Aeolid nudibranch, *Hermissenda crassicornis*
Sea hare, *Aplysia*

26 Chambered nautilus, *Nautilus*

27 Squid, *Loligo opalescens*
Octopus, *Octopus*

28 Acorn barnacle, *Balanus*
Copepod, *Calanus*
Stalked barnacle, *Lepas*
Amphipod, Generalized Gammaridian
Isopod, *Ligia pallasii*

29 Shrimp, *Heptacarpus*
Hermit crab, *Pagurus*
Sand crab, *Emerita*
Lobster, *Homarus americanus*

30 True crab, Tribe Brachyura
Cancer crab, *Cancer antennarius*
Shore crab, *Pachygrapsus crassipes*
Blue crab, *Callinectes sapidus*
Box crab, *Calappa*

31 Sea star, *Pisaster*

32 Brittle star, Class Ophiuroidea
Feather star, Class Crinoidea

33 Sea urchin, *Strongylocentrotus*
Sand dollar, *Dendraster excentricus*
Sea cucumber (deposit feeder), *Parastichopus*
Sea cucumber (filter feeder), *Cucumaria miniata*

34 Sea squirt (solitary), *Ciona intestinalis*
Sea grape, *Molgula manhattensis*
Compound tunicate, *Botryllus*
Larvacean tunicate, *Oikopleura*

35 Grouper, Family Serranidae

36 Flying fish, *Cypselurus*, Family Exocoetidae
Northern herring, *Clupea harengus*, Family Clupeidae
Swordfish, *Xiphias gladius*, Family Xiphiidae
Sunfish, *Mola mola*, Family Molidae
Albacore, *Thunnus alalunga*, Family Scombridae

37 Tidepool sculpin, *Oligocottus maculosus*, Family Cottidae
Sea robin, *Prionotus*, Family Triglidae

Stargazer, *Astroscopus*, Family Uranoscopidae
Starry flounder, *Platichtys stellatus*, Family Pleuronectidae

38 Trumpetfish, *Aulostomus chinensis*, Family Aulostomidae
Coral grouper, *Cephalopholis miniatus*, Family Serranidae
Golden boxfish, *Ostracion tuberculatus*, Family Ostraciontidae
Moray eel, *Gymnothorax*, Family Muraenidae
Butterflyfish, *Chaetodon auriga*, Family Chaetodontidae

39 Hatchetfish, *Argyropelecus*, Family Sternoptychidae
Lanternfish, *Myctophum*, Family Myctophidae
Pacific viperfish, *Chauliodus macouni*, Family Chauliodontidae
Black devil, *Melanocetus johnsonii*, Family Melanocetidae
Pelican gulper, *Eurypharynx pelecanoides*, Family Eurypharyngidae
Tripodfish, *Bathypterois viridensis*, Family Bathyteroidae

41 Spiny dogfish shark, *Squalus acanthias*, Family Squalidae
Skate, *Raja binoculata*, Family Rajidae
Stingray, *Dasyatis americana*, Family Dasyatididae

42 Basking shark, *Cetorhinus maximus*, Family Cetorhinidae
Hammerhead shark, *Sphyrna mokarran*, Family Sphyrnidae
Great white shark, *Carcharodon carcharias*, Family Lamnidae
Thresher shark, *Alopias vulpinus*, Family Alopidae

43 Manta ray, *Manta birostris*, Family Mobulidae
Spotted eagle ray, *Aetobatus narinari*, Family Myliobatidae
Sawfish, *Pristis*, Family Pristidae
Electric ray, *Torpedo nobiliana* Family Torpedinidae

44 Green sea turtle, *Chelonia mydas*
Yellow-bellied sea snake, *Pelamis platurus*

45 Sea otter, *Enhydra lutris*, Family Mustelidae
California sea lion, *Zalophus californianus*, Order Pinnipedia
Dugong, *Dugong dugon*, Order Sirenia
Bottlenose dolphin, *Tursiops truncatus*, Order Cetacea

46 Fur seal, *Callorhinus ursinus*, Family Otariidae
Harbor seal, *Phoca vitulina*, Family Phocidae
Walrus, *Odobenus rosmarus*, Family Odobenidae

Elephant seal, *Mirounga angustirostris*, Family Phocidae

47 Toothed whales, Suborder Odontoceti
Dolphin, *Tursiops*
Sperm whale, *Physeter catodon*
Blue whale, *Balaenoptera musculus*
Killer whale, *Orcinus orca*

48 Baleen whales, Suborder Mysticeti
Right whale, *Eubalaena glacialis*
Humpback whale, *Megatera novaeangliae*
Gray whale, *Eschrictius robustus*

49 Garibaldi, *Hypsypops rubicundus*, Family Pomacentridae
Lionfish, *Pterois lunulata*, Family Scorpaenidae
Koran angelfish, *Pomacanthus semicirculatus*, Family Chaetodontidae
Copperband butterflyfish, *Chelmon rostratus*, Family Chaetodontidae

50 Common mackerel, *Scomber scomber*, Family Scombridae
Grouper, *Epinephelus*, Family Serranidae
Clownfish, *Amphiprion tricinctus*, Family Pomacentridae
Stonefish, *Synanceia horrida*, Family Scorpaenidae

52 Peppermint shrimp, *Hippolysmata grabbami*
Cuttlefish, *Sepia officinalis*
Nudibranch, *Chromodoris*

53 Red sponge nudibranch, *Rostanga pulchra*
Red sponge, *Ophlitaspongia pennata*
Isopod, *Idothea Montereyensis*
Red alga, *Plocamium*
Limpet, *Collisella digitalis*
Stalked barnacle, *Pollicipes polymerus*
Decorator crab, *Pugettia richii*

54 Fiddler crab, *Uca*

55 Dinoflagellate, *Noctiluca*
Fireworm, *Odontosyllis enopla*
Sea gooseberry, *Pleurobrachia*
Whale krill, *Euphausia*
Firefly squid, *Watasenia scintillans*

56 Lanternfish, *Myctophum*, Family Myctophidae
Hatchetfish, *Argyropelecus*, Family Sternopthychidae
Stomiatoid, *Idiacanthus*, Family Stomiatoidae
Flashlight fish, *Photoblepharon*, Family Anomalopidae

57 Armored dinoflagellate, *Gonyaulax*

58 Green alga, *Ulva*
Bull kelp, *Nereocystis luetkeana*
Nori, *Porphyra*

59 Turtle grass, *Thallassia testudinum*
Sea star, Class Asteroidea
Sea anemone, *Metridium*
Coral, Class Anthozoa
Polychaete worm, *Autolytus prolifer*

60 Brittle star, Class Ophiuroidea
Porcelain crab, *Petrolisthes*
Sunfish, *Mola mola,* Family Molidae
Polychaete worm, *Nereis*

61 Hydrozoan polyp colony, *Obelia*
Jellyfish, *Aurelia*
Sea anemone, *Epiactis prolifera*

62 Polychaete worm, *Spirorbis*
Clam worm, *Nereis*
Palolo worm, *Palola viridis*

63 Virginia oyster, *Crassostrea virginica*

64 Abalone, *Haliotis*
Moon snail, *Polinices*
Whelk, *Nucella emarginata*
Dorid nudibranch, *Archidoris*

65 Octopus, *Octopus vulgaris*
Squid, *Loligo opalescens*
Paper nautilus, *Argonauta argo*

66 Barnacle, *Balanus*
Copepod, *Calanus*

67 Generalized gammaridian Amphipod, *Caprella,*
Mantis shrimp (stomatopod), *Squilla*
Spiny lobster, *Panulirus interruptus*
Shrimp, *Heptacarpus*

68 Red rock crab, *Cancer productus*

69 Sea urchin, Class Echinoidea
Six-rayed sea star, *Leptasterias hexactis*

70 Horn shark, *Heterodontus,* Family Heterodontidae
Spotted dogfish shark, *Scyliorhinus caniculus,* Family Scyliorhinidae
Skate, *Raja binoculata,* Family Rajidae
Smoothhound shark, *Mustelus manazo,* Family Carcharhinidae

71 Surfperch, *Amphistichus,* Family Embiotocidae
Gafftopsail catfish, *Barge marinus,* Family Ariidae
Kurtus, *Kurtus,* Family Kurtidae
Sea horse, *Hippocampus,* Family Syngnathidae

72 Sockeye salmon, *Oncorhynchus nerka,* Family Salmonidae
Grunion, *Leuresthes tenuis, Family Atherinidae*
Damselfish, Pomacentrus, Family Pomacentridae

73 Northern herring, *Clupea harengus,* Family Clupeidae
European eel, *Anguilla anguilla,* Family Anguillidae

74 California gray whale, *Eschrichtius robustus*

75 Elephant seal, *Mirounga angustirostris*

76 Sea slug, *Elysia viridis*
Green alga, *Codium*
Giant clam, *Tridacna*
Green sea anemone, *Anthopleura xanthogrammica*
Elkhorn coral, *Acropora palmata*
Brain coral, *Colophyllia natans*

77 Cleaner shrimp, *Periclimenes pedersoni*
Sea anemone, *Bartholomea annulata*
Cleaner wrasse, *Labroides dimidiatus,* Family Labridae
False cleaner blenny, *Aspidontus taeniatus,* Family Blennidae

78 Anemonefish, *Amphiprion,* Family Pomacentridae
Butterflyfish, *Chaetodon lunula,* Family Chaetodontidae

79 Aggregating anemone, *Anthopleura elegantissima*

81 Limpet, *Lottia gigantea*
Mussel, *Mytilus californianus*
Stalked barnacle, *Pollicipes polymerus*
Snail, *Nucella emarginata*

82 Sea palm, *Postelsia palmaeformis*
California sea mussel, *Mytilus californianus*
Sea star, *Pisaster ochraceus*
Barnacle, *Balanus cariosus*

83 Aeolid nudibranch, *Aeolidia papillosa*
Aggregating anemone, *Anthopleura elegantissima*

84 Ochre star, *Pisaster ochraceus*
Keyhole limpet, *Diodora aspera*
Leather star, *Dermasterias imbricata*
Sea urchin, *Strongylocentrotus purpuratus*
Feather-duster worm, *Eudistylia*

85 Porcupinefish, *Diodon hystrix,* Family Diodontidae
Surgeonfish, *Acanthurus glaucopareius,* Family Acanthuridae
Clown triggerfish, *Balistoides niger,* Family Balistidae
Shrimpfish, *Aeoliscus strigatus,* Family Centriscidae

86 Scallop, *Pecten*
Short-spined star, *Pisaster brevispinus*
Cockle, *Clinocardium nuttalli*
Sea anemone, *Stomphia coccinea*
Leather star, *Dermasterias imbricata*

87 Common octopus, *Octopus vulgaris*
Pelagic octopus, *Tremoctopus*

88 Pacific sand dollar, *Dendraster excentricus*
Polychaete worm, *Chaetopterus variopedatus*

89 Ringed-top shell snail, *Calliostoma annulatum*
Abalone, *Haliotis*
Oyster drill, *Urosalpinx cinera*

90 Sea urchin, *Strongylocentrotus*
Giant kelp, *Macrocystis*
Beach hoppers, *Orchestoidea*
Limpet, *Notoacmea insessa*
Feather-boa kelp, *Egregia menziesii*
Lined chiton, *Tonicella lineata*

91 Cone snail, *Conus*
Mantis shrimp, order Stomatopoda
Sea anemone, *Tealia*
Bat star, *Patiria miniata*
Crown-of-thorns sea star, *Acanthaster planci*

92 Great barracuda, *Sphyraena barracuda,* Family Sphyraenidae
Bluefish, *Pomatomus salatrix,* Family Pomatomidae
Greater amberjack, *Seriola dumerili,* Family Carangidae
Splitlure frogfish, *Antennarius scaber,* Family Anternnariidae

93 California sheephead, *Semicossyphus pulcher,* Family Labridae
Pipefish, *Syngnathus,* Family Syngnathidae
Reef filefish, *Oxymonacanthus longirostris,* Family Monacanthidae
Forceps butterflyfish, *Forcipiger flavissimus,* Family Chaetodontidae
Queen triggerfish, *Balistes vetula,* Family Balistidae
Sea urchin, *Diadema antillarum*

94 Northern anchovy, *Engraulis mordax,* Family Engraulidae
Parrotfish, *Scarus,* Family Scaridae
Trunkfish, *Ostracion,* Family Ostraciontidae
Bat stingray, *Myliobatus californicus,* Family Myliobatidae
Goatfish, *Mulloidichthys,* Family Mullidae

95 Sea star, *Pisaster ochraceus*
Mussel, *Mytilus californianus*
Feather-boa kelp, *Egregia menziesii*
Sea palm, *Postelsia palmaeformis*
Stalked barnacle, *Pollicipes polymerus*
Green sea anemone, *Anthopleura xanthogrammica*
Chiton, *Katharina tunicata*
Red alga, *Endocladia muricata*

96 Sea otter, *Enhydra lutris*
Sea urchin, *Strongylocentrotus*
Abalone, *Haliotis*

INDEX